On the Shoulders of Giants

Selections from *The Principle of Relativity*

On the Shoulders of Giants

Selections from *The Principle of Relativity*

by Albert Einstein
Edited, with Commentary, by Stephen Hawking

RUNNING PRESS
PHILADELPHIA • LONDON

Printed in the United States

9 8 7 6 5 4 3

Digit on the right indicates the number of this printing

Library of Congress Control Number: 2004094109

ISBN 0-7624-2023-5

Author photo courtesy of Book Laboratory

Cover Design: Doogie Horner

Typography: Scribe

Cover images: Portrait of Albert Einstein: Getty Images; (bottom) Eskimo Nebula STScI-2000-07, courtesy of NASA, Andrew Fruchter and the ERO Team [Sylvia Baggett (STScI), Richard Hook (ST-ECF), Zoltan Levay (STScI)]

Selections from *The Principle of Relativity: A Collection of Papers on the Special and General Theory of Relativity*, courtesy of Dover Publications.

This book may be ordered by mail from the publisher. Please include $2.50 for postage and handling.

But try your bookstore first!

Running Press Book Publishers

125 South Twenty-second Street

Philadelphia, PA 19103-4399

Visit us on the web!

www.runningpress.com

Contents

A Note to the Reader

Selections from The Principle of Relativity is one volume in a series of five scientific classics that Running Press has chosen to publish in a new format with commentary by renowned physicist, Stephen Hawking. Hawking chose these five particular essays to illuminate the evolution of modern physics and astronomy, as well as the process through which scientific knowledge evolves.

Also included in this series are:

- *On the Revolutions of Heavenly Spheres* by Nicolaus Copernicus
- *Dialogues Concerning Two New Sciences* by Galileo Galilei
- *Harmonies of the World* by Johannes Kepler
- *Principia* by Sir Isaac Newton

Look for all five essays combined into one volume in:

- *On the Shoulders of Giants* edited, with commentary, by Stephen Hawking
- *The Illustrated On the Shoulders of Giants* edited, with commentary, by Stephen Hawking

The text for *The Principle of Relativity* is based on a translation of the original, printed edition. We have made no attempt to modernize the author's own distinct usage, spelling, or punctuation. The details of this essay are as follows:

We have chosen seven works by Albert Einstein from *The Principle of Relativity: A Collection of Original Papers on the Special Theory of Relativity*, by H.A. Lorentz, A. Einstein, H. Minkowski and H. Weyl. The entire collection was originally published in German, under the title "*Des Relativitatsprinzip*" in 1922. Our text comes from the translation by W. Perrett and G.B. Jeffrey.

The Editors

Introduction

If I have seen farther, it is by standing on the shoulders of giants, wrote Isaac Newton in a letter to Robert Hooke in 1676. Although Newton was referring to his discoveries in optics rather than his more important work on gravity and the laws of motion, it is an apt comment on how science, and indeed the whole of civilization, is a series of incremental advances, each building on what went before. This is the theme of this fascinating series, which uses the original texts to trace the evolution of our picture of the heavens from the revolutionary claim of Nicolaus Copernicus that the earth orbits the sun to the equally revolutionary proposal of Albert Einstein that space and time are curved and warped by mass and energy. It is a compelling story because both Copernicus and Einstein have brought about profound changes in what we see as our position in the order of things. Gone is our privileged place at the center of the universe, gone are eternity and certainty, and gone are Absolute Space and Time to be replaced by rubber sheets.

It is no wonder both theories encountered violent opposition: the Inquisition in the case of the Copernican theory and the Nazis in the case of Relativity. We now have a tendency to dismiss as primitive the earlier world picture of Aristotle and Ptolemy in which the Earth was at the center and the Sun went round it. However we should not be too scornful of their model, which was anything but simpleminded. It incorporated Aristotle's deduction that the Earth is a round ball rather than a flat plate and it was reasonably accurate in its main function, that of predicting the apparent positions of the heavenly bodies in the sky for astrological purposes. In fact, it was about as accurate as the heretical suggestion put forward in 1543 by Copernicus that the Earth and the planets moved in circular orbits around the Sun.

Galileo found Copernicus's proposal convincing not because it better fit the observations of planetary positions but because of its simplicity and elegance, in contrast to the complicated epicycles of the Ptolemaic model. In *Dialogues Concerning Two New Sciences,* Galileo's characters, Salviati and Sagredo, put forward persuasive arguments in support of Copernicus. Yet, it was still possible for his third character, Simplicio, to defend Aristotle and Ptolemy and to maintain that in reality the Earth was at rest and the Sun went round the Earth.

It was not until Kepler's work made the Sun-centered model more accurate and Newton gave it laws of motion that the Earth-centered picture finally lost all credibility. It was quite a shift in our view of the universe: If we are not at the center, is our existence of any importance? Why should God or the Laws of Nature care about what happens on the third rock from the Sun, which is where Copernicus has left us? Modern scientists have out-Copernicused Copernicus by seeking an account of the universe in which Man (in the old pre-politically correct sense) played no role. Although this approach has succeeded in finding objective impersonal laws that govern the universe, it has not (so far at least) explained why the universe is the way it is rather than being one of the many other possible universes that would also be consistent with the laws.

Some scientists would claim that this failure is only provisional, that when we find the ultimate unified theory, it will uniquely prescribe the state of the universe, the strength of gravity, the mass and charge of the electron and so on. However, many features of the universe (like the fact that we are on the third rock, rather than the second or fourth) seem arbitrary and accidental and not the predictions of a master equation. Many people (myself included) feel that the appearance of such a complex and structured universe from simple laws requires the invocation of something called the anthropic principle, which restores us to the central position we have been too modest to claim since the time of Copernicus. The anthropic principle is based on the self-evident fact that we wouldn't be asking questions about the nature of the universe if the universe hadn't contained stars, planets and stable chemical compounds, among other prerequisites of (intelligent?) life as we know it. If the ultimate theory made a unique prediction for the state of the universe and its contents, it would be a remarkable coincidence that this state was in the small subset that allow life.

However the work of the last thinker in this series, Albert Einstein, raises a new possibility. Einstein played an important role in the development of quantum theory which says that a system doesn't just have a single history as one might have thought. Rather it has every possible history with some probability. Einstein was also almost solely responsible for the general theory of relativity in which space and time are curved and become dynamic. This means that they are subject to quantum theory and that the universe itself has every possible shape and history. Most of these histories will be quite unsuitable for the development of life but a very few have all the conditions needed. It doesn't matter if these few have a very low probability relative to the others: the lifeless universes will have no one to observe them. It is sufficient that there is at least one history in which life develops, and we ourselves are evidence for that, though maybe not for intelligence. Newton said he was "*standing on the shoulders of giants.*" But as this series illustrates so well, our understanding doesn't advance just by slow and steady building on previous work. Sometimes as with Copernicus and Einstein, we have to make the intellectual leap to a new world picture. Maybe Newton should have said, "*I used the shoulders of giants as a springboard.*"

Albert Einstein

(1879–1955)
HIS LIFE AND WORK

Genius isn't always immediately recognized. Although Albert Einstein would become the greatest theoretical physicist who ever lived, when he was in grade school in Germany his headmaster told his father, "He'll never make a success of anything." When Einstein was in his mid-twenties, he couldn't find a decent teaching job even though he had graduated from the Federal Polytechnic School in Zurich as a teacher of mathematics and physics. So he gave up hope of obtaining a university position and applied for temporary work in Bern. With the help of a classmate's father, Einstein managed to secure civil-service post as an examiner in the Swiss patent office. He worked six days a week, earning $600 a year. That's how he supported himself while working toward his doctorate in physics at the University of Zurich.

In 1903, Einstein married his Serbian sweetheart, Mileva Maric, and the couple moved into a one-bedroom flat in Bern. Two years later, she bore him a son, Hans Albert. The period surrounding Hans's birth was probably the happiest time in Einstein's life. Neighbors later recalled seeing young father absentmindedly pushing a baby carriage down the city streets. From time to time, Einstein would reach into the carriage and remove a pad of paper on which to jot downnotes to himself. It seems likely that the notepad in the baby's stroller contained some of the formulas and equations that led to the theory of relativity andthe development of the atomic bomb.

During these early years at the patent office, Einstein spent most of his spare time studying theoretical physics. He composed a series of four seminal scientific papers which set forth some of the most momentous ideas in the long history of the quest to comprehend the universe. Space and time would never be looked at the same way again. Einstein's work won him the Nobel Prize in Physics in 1921, as well as much popular acclaim.

As Einstein pondered the workings of the universe, he received flashes of understanding that were too deep for words. "These thoughts did not come in any verbal formulation," Einstein was once quoted as saying. "I rarely think in words at all. A thought comes, and I may try to express it in words afterward."

Einstein eventually settled in the United States, where he publicly championed such causes as Zionism and nuclear disarmament. But he maintained his passion for physics. Right up until his death in 1955, Einstein kept seeking a unified field theory that would link the phenomena of gravitation and electromagnetism in one set of equations. It is a tribute to Einstein's vision that physicists today continue to seek a grand unification of physical theory. Einstein revolutionized scientific thinking in the twentieth century and beyond.

Albert Einstein was born at Ulm, in the former German state of Wüttemberg on March 14, 1879, and grew up in Munich. He was the only son of Hermann Einstein and Pauline Koch. His father and uncle owned an electrotechnical plant. The family considered Albert a slowlearner because he had difficulty with language. (It is now thought that he may have been dyslexic.) Legend has it that when Hermann asked the headmaster of his son's school about the best profession for Albert, the man replied, "It doesn't matter. He'll never make a success of anything."

Einstein did not do well in school. He didn't like the regimentation, and he suffered from being one of the few Jewish children in a Catholic school. This experience as an outsider was one that would repeat itself many times in his life.

One of Einstein's early loves was science. He remembered his father showing him a pocket compass when he was around five years old, and marveling that the needle always pointed north, even if the case was spun. In that moment, Einstein recalled, he "felt something deeply hidden had to be behind things."

Another of his early loves was music. Around the age of six, Einstein began studying the violin. It did not come naturally to him but when after several years he recognized the mathematical structure of music, the violin became a lifelong passion—although his talent was never a match for his enthusiasm.

When Einstein was ten his family enrolled him in the Luitpold Gymnasium, which is where, according to scholars, he developed a suspicion of authority. This trait served Einstein well later in life as a scientist. His habit of skepticism made it easy for him to question many long-standing scientific assumptions.

In 1895, Einstein attempted to skip high school by passing an entrance examinationto the Federal Polytechnic School (ETH) in Zurich, where he hoped to pursue a degree in electrical engineering. This is what he wrote about his ambitions at the time:

> *If I were to have the good fortune to pass my examinations, I would go to Zurich. I would stay there for four years in order to study mathematics and physics. I imagine myself becoming a teacher in those branches of the natural sciences, choosing the theoretical part of them. Here are the reasons which lead me to this plan. Above all, it is my disposition for abstract and mathematical thought, and my lack of imagination and practical ability.*

Einstein failed the arts portion of the exam so was denied admission to the polytechnic. His family instead sent him to secondary school at Aarau, in Switzerland, hoping that it would earn him for a second chance to enterthe Zurich school. It did, and Einstein graduated from the polytechnic in 1900. At about that time he fell in love with Mileva Maric, and in 1901 she gave birth, out of wedlock to their first child, a daughter named Lieserl. Very little is known for certain about Lieserl, but it appears that she either was born with a crippling condition or fell very ill as an infant then was put up for adoption, and died at about two years of age. Einstein and Maric married in 1903.

The year Hans was born, 1905, was a miracle year for Einstein. Somehow he managed to handle the demands of fatherhood and a full-time job and still publish four epochal scientific papers all without benefit of the resources that an academic appointment might have provided.

In the spring of that year, Einstein submitted three papers to the German periodical *Annals of Physics* (*Annalen der Physik*). The three appeared together in the journal's volume 17. Einstein characterized the first paper, on the light quantum, as "very revolutionary." In it, he examined the phenomenon of the quantum (the fundamental unit of energy) discovered by the German physicist Max Planck. Einstein explained the photoelectric effect, which holds that for each electron emitted, a specific amount of energy is released. This is the quantum effect that states that energy is emitted in fixed amounts that can be expressed only as whole integers. This theory formed the basis for a great deal of quantum mechanics. Einstein suggested that light be considered a collection of independent particles of energy, but remarkably, he offered no experimental data. He simply argued hypothetically for the existence of these "light quantum" for aesthetic reasons.

Initially, physicists were hesitant to endorse Einstein's theory. It was too great a departure from scientifically accepted ideas of the time, and far beyond anything Planck had discovered. It was this first paper, titled "On a Heuristic View concerning the Production and Transformation of Light"—not his work on relativity—that won Einstein the Nobel Prize in Physics in 1921.

In his second paper, "On a New Determination of Molecular Dimensions"—which Einstein wrote as his doctoral dissertation—and his third, "On the Movement of Small Particles Suspended in Stationary Liquids Required by the Molecular-Kinetic Theory of Heat," Einstein proposed a method to determine the size and motion of atoms. He also explained Brownian motion, a phenomenon described by the British botanist Robert Brown after studying the erratic movement of pollen suspended in fluid. Einstein asserted that this movement was caused by impacts between atoms and molecules. At the time, the very existence of atoms was still a subject of scientific debate, so there could be no underestimating the importance of these two papers. Einstein had confirmed the atomic theory of matter.

In the last of his 1905 papers, entitled "On the Electrodynamics of Moving Bodies" Einstein presented what became known as the special theory of relativity. The paper reads more like an essay than a scientific communication. Entirely theoretical, it contains no notes or bibliographic citations. Einstein wrote this 9,000-word treatise in just five weeks yet historians of science consider it every bit as comprehensive and revolutionary as Isaac Newton's *Principia.*

What Newton had done for our understanding of gravity, Einstein had done for our view of time and space, managing in the process to overthrow the Newtonian conception of time. Newton had declared that "absolute, true, and mathematical time, of itself and from its own nature, flows equably without relation to anything external." Einstein held that all observers should measure the same speed for light, regardless of how fast they themselves are moving. Einstein also asserted that the mass of an object is not unchangeable but rather increases with the object's velocity. Experiments later proved that a small particle of matter, when accelerated to 86 percent of the speed of light, has twice as much mass as it does at rest.

Another consequence of relativity is that the relation between energy and mass may be expressed mathematically, which Einstein did in the famous equation $E=mc^2$. This expression—that energy is equivalent to mass times the square of the speed of light—led physicists to understand that even miniscule amounts of matter have the potential to yield enormous amounts of energy. Completely converting to energy just a part of the mass of a few atoms would, then, result in a colossal explosion. Thus did Einstein's modest-looking equation lead scientists to consider the consequences of splitting the atom (nuclear fission) and, at the urging of governments, to develop the atomic bomb. In 1909, Einstein was appointed professor of theoretical physics at the University of Zurich and three years later he fulfilled his ambition to return to the Federal Polytechnic School as a full professor. Other prestigious academic appointments and directorships followed. Throughout, he continued to work on his theory of gravity as well as his general theory of relativity. But as his professional status continued to rise, his marriage and health began to deteriorate. He and Mileva began divorce proceedings in 1914, the same year he accepted a professorship at the University of Berlin. When he later fell ill, his cousin Elsa nursed him back to health and around 1919 they were married.

Where the special theory of relativity radically altered concepts of time and mass, the general theory of relativity changed our concept of space. Newton had written that "absolute space, in its own nature, without relation to anything external, remains always similar and immovable." Newtonian space is Euclidean, infinite, and unbounded. Its geometric structure is completely independent of the physical matter occupying it. In it, all bodies gravitate toward one another without having any effect on the structure of space. In stark contrast, Einstein's general theory of relativity asserts that not only does a body's gravitational mass act on other bodies, it also influences the structure of space. If a body is massive enough, it induces space to curve around it. In such a region, light appears to bend.

In 1919, Sir Arthur Eddington sought evidence to test the general theory. Eddington organized two expeditions—one to Brazil and the other to West Africa—to observe the light from stars as it passed near a massive body—the sun—during a total solar eclipse on May 29. Under normal circumstances such observations would be impossible, as the weak light from distant stars would be blotted out by daylight, but during the eclipse such light would briefly be visible.

In September, Einstein received a telegram from Hendrik Lorentz, a fellow physicist and close friend. It read: "Eddington found star displacement at rim of Sun, preliminary measurements between nine-tenths of a second and twice that value." Eddington's data were in keeping with the displacement predicted by the special relativity theory. His photographs from Brazil seemed to show the light from known stars in a different position in the sky during the eclipse than they were at nighttime, when their light did not pass near the sun. The theory of general relativity had been confirmed, forever changing the course of physics. Years later, when a student of Einstein's asked how he would have reacted had the theory been disproved, Einstein replied, "Then I would have felt sorry for the dear Lord. The theory is correct."

Confirmation of general relativity made Einstein world-famous. In 1921, he was elected a member of the British Royal Society. Honorary degrees and awards greeted him at every city he visited. In 1927, he began developing the foundation of quantum mechanics with the Danish physicist Niels Bohr, even as he continued to pursue his dream of a unified field theory. His travels in the United States led to his appointment in 1932 as a professor of mathematics and theoretical physics at the Institute for Advanced Study in Princeton, New Jersey.

A year later, he settled permanently in Princeton after the ruling Nazi party in Germany began a campaign against "Jewish science." Einstein's property was confiscated, and he was deprived of German citizenship and positions in German universities. Until then, Einstein had considered himself a pacifist. But when Hitler turned Germany into a military power in Europe, Einstein came to believe that the use of force against Germany was justified. In 1939, at the dawn of World War II, Einstein became concerned that the Germans might be developing the capability to build an atomic bomb—a weapon made possible by his own research and for which he therefore felt a responsibility. He sent a letter to President Franklin D. Roosevelt warning of such a possibility and urging that the United States undertake nuclear research. The letter, composed by his friend and fellow scientist Leo Szilard, became the impetus for the formation of the Manhattan Project, which produced the world's first atomic weapons. In 1944, Einstein put a handwritten copy of his 1905 paper on special relativity up for auction and donated the proceeds—six million dollars—to the Allied war effort. After the war, Einstein continued to involve himself with causes and issues that concerned him. In November 1952, having shown strong support for Zionism for many years, he was asked to accept the presidency of Israel. He respectfully declined, saying that he was not suited for the position. In April 1955, only one week before his death, Einstein composed a letter to the philosopher Bertrand Russell in which he agreed to sign his name to a manifesto urging all nations to abandon nuclear weapons.

Einstein died of heart failure on April 18, 1955. Throughout his life, he had sought to understand the mysteries of the cosmos by probing it with his thought rather than relying on his senses. "The truth of a theory is in your mind," he once said, "not in your eyes."

THE PRINCIPLE OF RELATIVITY

Translated by W. Perrett and G. B. Jeffery

ON THE ELECTRODYNAMICS OF MOVING BODIES

It is known that Maxwell's electrodynamics—as usually understood at the present time—when applied to moving bodies, leads to asymmetries which do not appear to be inherent in the phenomena. Take, for example, the reciprocal electrodynamic action of a magnet and a conductor. The observable phenomenon here depends only on the relative motion of the conductor and the magnet, whereas the customary view draws a sharp distinction between the two cases in which either the one or the other of these bodies is in motion. For if the magnet is in motion and the conductor at rest, there arises in the neighbourhood of the magnet an electric field with a certain definite energy, producing a current at the places where parts of the conductor are situated. But if the magnet is stationary and the conductor in motion, no electric field arises in the neighbourhood of the magnet. In the conductor, however, we find an electromotive force, to which in itself there is no corresponding energy, but which gives rise—assuming equality of relative motion in the two cases discussed—to electric currents of the same path and intensity as those produced by the electric form in the former case.

Examples of this sort, together with the unsuccessful attempts to discover any motion of the earth relatively to the "light medium," suggest that the phenomena of electrodynamics as well as of mechanics possess no properties corresponding to the idea of absolute rest. They suggest rather that, as has already been shown to the first order of small quantities, the same laws of electrodynamics and optics will be valid for all frames of reference for which the equations of mechanics hold good.* We will raise this conjecture (the purport of which will hereafter be called the "Principle of Relativity") to the status of a postulate, and also introduce another postulate, which is only apparently

* The preceding memoir by Lorentz was not at this time known to the author.

irreconcilable with the former, namely, that light is always propagated in empty space with a definite velocity *c* which is independent of the state of motion of the emitting body. These two postulates suffice for the attainment of a simple and consistent theory of the electrodynamics of moving bodies based on Maxwell's theory for stationary bodies. The introduction of a "luminiferous ether" will prove to be superfluous inasmuch as the view here to be developed will not require an "absolutely stationary space" provided with special properties, nor assign a velocity-vector to a point of the empty space in which electromagnetic processes take place.

The theory to be developed is based—like all electrodynamics—on the kinematics of the rigid body, since the assertions of any such theory have to do with the relationships between rigid bodies (systems of co-ordinates), clocks, and electromagnetic processes. Insufficient consideration of this circumstance lies at the root of the difficulties which the electrodynamics of moving bodies at present encounters.

I. KINEMATICAL PART

§ 1. DEFINITION OF SIMULTANEITY

Let us take a system of co-ordinates in which the equations of Newtonian mechanics hold good.* In order to render our presentation more precise and to distinguish this system of co-ordinates verbally from others which will be introduced hereafter, we call it the "stationary system."

If a material point is at rest relatively to this system of co-ordinates, its position can be defined relatively thereto by the employment of rigid standards of measurement and the methods of Euclidean geometry, and can be expressed in Cartesian co-ordinates.

If we wish to describe the *motion* of a material point, we give the values of its co-ordinates as functions of the time. Now we must bear carefully in mind that a mathematical description of this kind has no physical meaning unless we are quite clear as to what we understand by "time." We have to take into account that all our judgments in which time plays a part are always judgments of *simultaneous events.* If, for instance, I say, "That train arrives here at 7 o'clock," I mean something like this: "The pointing of the small hand of my watch to 7 and the arrival of the train are simultaneous events."†

It might appear possible to overcome all the difficulties attending the definition of "time" by substituting "the position of the small hand of my watch" for "time." And in fact such a definition is satisfactory when we are concerned with defining a time

*I.e. to the first approximation.

†We shall not here discuss the inexactitude which lurks in the concept of simultaneity of two events at approximately the same place, which can only be removed by an abstraction.

exclusively for the place where the watch is located; but it is no longer satisfactory when we have to connect in time series of events occurring at different places, or—what comes to the same thing—to evaluate the times of events occurring at places remote from the watch.

We might, of course, content ourselves with time values determined by an observer stationed together with the watch at the origin of the co-ordinates, and co-ordinating the corresponding positions of the hands with light signals, given out by every event to be timed, and reaching him through empty space. But this co-ordination has the disadvantage that it is not independent of the standpoint of the observer with the watch or clock, as we know from experience. We arrive at a much more practical determination along the following line of thought.

If at the point A of space there is a clock, an observer at A can determine the time values of events in the immediate proximity of A by finding the positions of the hands which are simultaneous with these events. If there is at the point B of space another clock in all respects resembling the one at A, it is possible for an observer at B to determine the time values of events in the immediate neighbourhood of B. But it is not possible without further assumption to compare, in respect of time, an event at A with an event at B. We have so far defined only an "A time" and a "B time." We have not defined a common "time" for A and B, for the latter cannot be defined at all unless we establish *by definition* that the "time" required by light to travel from A to B equals the "time" it requires to travel from B to A. Let a ray of light start at the "A time" t_A from A towards B, let it at the "B time" t_B be reflected at B in the direction of A, and arrive again at A at the "A time" t'_A.

In accordance with definition the two clocks synchronize if

$$t_B - t_A = t'_A - t_B.$$

We assume that this definition of synchronism is free from contradictions, and possible for any number of points; and that the following relations are universally valid:

1. If the clock at B synchronizes with the clock at A, the clock at A synchronizes with the clock at B.

2. If the clock at A synchronizes with the clock at B and also with the clock at C, the clocks at B and C also synchronize with each other.

Thus with the help of certain imaginary physical experiments we have settled what is to be understood by synchronous stationary clocks located at different places, and have evidently obtained a definition of "simultaneous" or "synchronous," and of "time." The "time" of an event is that which is given simultaneously with the event by a stationary clock located at the place of the event, this clock being synchronous, and indeed synchronous for all time determinations, with a specified stationary clock.

In agreement with experience we further assume the quantity

$$\frac{2AB}{t'_A - t_A} = c$$

to be a universal constant—the velocity of light in empty space.

It is essential to have time defined by means of stationary clocks in the stationary system, and the time now defined being appropriate to the stationary system we call it "the time of the stationary system."

§ 2. ON THE RELATIVITY OF LENGTHS AND TIMES

The following reflexions are based on the principle of relativity and on the principle of the constancy of the velocity of light. These two principles we define as follows:—

1. The laws by which the states of physical systems undergo change are not affected, whether these changes of state be referred to the one or the other of two systems of co-ordinates in uniform translatory motion.

2. Any ray of light moves in the "stationary" system of co-ordinates with the determined velocity c, whether the ray be emitted by a stationary or by a moving body. Hence

$$\text{velocity} = \frac{\text{light path}}{\text{time interval}}$$

where time interval is to be taken in the sense of the definition in § 1.

Let there be given a stationary rigid rod; and let its length be l as measured by a measuring-rod which is also stationary. We now imagine the axis of the rod lying along the axis of x of the stationary system of co-ordinates, and that a uniform motion of parallel translation with velocity v along the axis of x in the direction of increasing x is then imparted to the rod. We now inquire as to the length of the moving rod, and imagine its length to be ascertained by the following two operations:—

(*a*) The observer moves together with the given measuring-rod and the rod to be measured, and measures the length of the rod directly by superposing the measuring-rod, in just the same way as if all three were at rest.

(*b*) By means of stationary clocks set up in the stationary system and synchronizing in accordance with § 1, the observer ascertains at what points of the stationary system the two ends of the rod to be measured are located at a definite time. The distance between these two points, measured by the measuring-rod already employed, which in this case is at rest, is also a length which may be designated "the length of the rod."

In accordance with the principle of relativity the length to be discovered by the operation (*a*)—we will call it "the length of the rod in the moving system"—must be equal to the length l of the stationary rod.

The length to be discovered by the operation (*b*) we will call "the length of the (moving) rod in the stationary system." This we shall determine on the basis of our two principles, and we shall find that it differs from *l*.

Current kinematics tacitly assumes that the lengths determined by these two operations are precisely equal, or in other words, that a moving rigid body at the epoch *t* may in geometrical respects be perfectly represented by *the same* body *at rest* in a definite position.

We imagine further that at the two ends A and B of the rod, clocks are placed which synchronize with the clocks of the stationary system, that is to say that their indications correspond at any instant to the "time of the stationary system" at the places where they happen to be. These clocks are therefore "synchronous in the stationary system."

We imagine further that with each clock there is a moving observer, and that these observers apply to both clocks the criterion established in § 1 for the synchronization of two clocks. Let a ray of light depart from A at the time* t_A, let it be reflected at B at the time t_B, and reach A again at the time t'_A. Taking into consideration the principle of the constancy of the velocity of light we find that

$$t_B - t_A = \frac{r_{AB}}{c - v} \quad \text{and} \quad t'_A - t_B = \frac{r_{AB}}{c + v}$$

where r_{AB} denotes the length of the moving rod—measured in the stationary system. Observers moving with the moving rod would thus find that the two clocks were not synchronous, while observers in the stationary system would declare the clocks to be synchronous.

So we see that we cannot attach any *absolute* signification to the concept of simultaneity, but that two events which, viewed from a system of co-ordinates, are simultaneous, can no longer be looked upon as simultaneous events when envisaged from a system which is in motion relatively to that system.

§ 3. THEORY OF THE TRANSFORMATION OF CO-ORDINATES AND TIMES FROM A STATIONARY SYSTEM TO ANOTHER SYSTEM IN UNIFORM MOTION OF TRANSLATION RELATIVELY TO THE FORMER

Let us in "stationary" space take two systems of co-ordinates, i.e. two systems, each of three rigid material lines, perpendicular to one another, and issuing from a point. Let the axes of X of the two systems coincide, and their axes of Y and Z respectively be parallel. Let each system be provided with a rigid measuring-rod and a number of clocks, and let the two measuring-rods, and likewise all the clocks of the two systems, be in all respects alike.

* "Time" here denotes "time of the stationary" and also "position of hands of the moving clock situated at the place, under discussion."

Now to the origin of one of the two systems (*k*) let a constant velocity *v* be imparted in the direction of the increasing *x* of the other stationary system (K), and let this velocity be communicated to the axes of the co-ordinates, the relevant measuring-rod, and the clocks. To any time of the stationary system K there then will correspond a definite position of the axes of the moving system, and from reasons of symmetry we are entitled to assume that the motion of *k* may be such that the axes of the moving system are at the time *t* (this "*t*" always denotes a time of the stationary system) parallel to the axes of the stationary system.

We now imagine space to be measured from the stationary system K by means of the stationary measuring-rod, and also from the moving system *k* by means of the measuring-rod moving with it; and that we thus obtain the co-ordinates x, y, z, and ξ, η, ζ respectively. Further, let the time t of the stationary system be determined for all points thereof at which there are clocks by means of light signals in the manner indicated in § 1; similarly let the time τ of the moving system be determined for all points of the moving system at which there are clocks at rest relatively to that system by applying the method, given in § 1, of light signals between the points at which the latter clocks are located.

To any system of values x, y, z, t, which completely defines the place and time of an event in the stationary system, there belongs a system of values ξ, η, ζ, τ, determining that event relatively to the system *k*, and our task is now to find the system of equations connecting these quantities.

In the first place it is clear that the equations must be *linear* on account of the properties of homogeneity which we attribute to space and time.

If we place $x' = x - vt$, it is clear that a point at rest in the system *k* must have a system of values x', y, z, independent of time. We first define τ as a function of x', y, z, and t. To do this we have to express in equations that τ is nothing else than the summary of the data of clocks at rest in system *k*, which have been synchronized according to the rule given in § 1.

From the origin of system *k* let a ray be emitted at the time τ_0 along the X-axis to x', and at the time τ_1 be reflected thence to the origin of the co-ordinates, arriving there at the time τ_2; we then must have $\frac{1}{2}(\tau_0 + \tau_2) = \tau_1$, or, by inserting the arguments of the function τ and applying the principle of the constancy of the velocity of light in the stationary system:

$$\frac{1}{2}\left[\tau(0,0,0,t) + \tau\left(0,0,0,t + \frac{x'}{c-v} + \frac{x'}{c+v}\right)\right] = \tau\left(x',0,0,t + \frac{x'}{c-v}\right)$$

Hence, if x' be chosen infinitesimally small,

$$\frac{1}{2}\left(\frac{1}{c-v} + \frac{1}{c+v}\right)\frac{\partial \tau}{\partial t} = \frac{\partial \tau}{\partial x'} + \frac{1}{c-v}\frac{\partial \tau}{\partial t}$$

or

$$\frac{\partial \tau}{\partial x'} + \frac{v}{c^2 - v^2}\frac{\partial \tau}{\partial t} = 0$$

It is to be noted that instead of the origin of the co-ordinates we might have chosen any other point for the point of origin of the ray, and the equation just obtained is therefore valid for all values of x', y, z.

An analogous consideration—applied to the axes of Y and Z—it being borne in mind that light is always propagated along these axes, when viewed from the stationary system, with the velocity $\sqrt{(c^2 - v^2)}$, gives us

$$\frac{\partial \tau}{\partial y} = 0, \frac{\partial \tau}{\partial z} = 0.$$

Since τ is a linear function, it follows from these equations that

$$\tau = a\left(t - \frac{v}{c^2 - v^2}x'\right)$$

where a is a function $\phi(v)$ at present unknown, and where for brevity it is assumed that at the origin of k, $\tau = 0$, when $t = 0$.

With the help of this result we easily determine the quantities ξ, η, ζ by expressing in equations that light (as required by the principle of the constancy of the velocity of light, in combination with the principle of relativity) is also propagated with velocity c when measured in the moving system. For a ray of light emitted at the time $\tau = 0$ in the direction of the increasing ξ

$$\xi = c\tau \text{ or } \xi = ac\left(t - \frac{v}{c^2 - v^2}x'\right).$$

But the ray moves relatively to the initial point of k, when measured in the stationary system, with the velocity $c - v$, so that

$$\frac{x'}{c - v} = t$$

If we insert this value of t in the equation for ξ, we obtain

$$\xi = a\frac{c^2}{c^2 - v^2}x'.$$

In an analogous manner we find, by considering rays moving along the two other axes, that

$$\eta = c\tau = ac\left(t - \frac{v}{c^2 - v^2}x'\right)$$

when

$$\frac{y}{\sqrt{(c^2 - v^2)}} = t, x' = 0$$

Thus

$$\eta = a\frac{c}{\sqrt{\left(c^2 - v^2\right)}}y \quad \text{and} \quad \zeta = a\frac{c}{\sqrt{\left(c^2 - v^2\right)}}z$$

Substituting for x' its value, we obtain

$$\tau = \phi(v)\beta\left(t - vx / c^2\right),$$
$$\xi = \phi(v)\beta(x - vt),$$
$$\eta = \phi(v)y,$$
$$\zeta = \phi(v)z$$

where

$$\beta = \frac{1}{\sqrt{\left(1 - v^2 / c^2\right)}},$$

and ϕ is an as yet unknown function of v. If no assumption whatever be made as to the initial position of the moving system and as to the zero point of τ, an additive constant is to be placed on the right side of each of these equations.

We now have to prove that any ray of light, measured in the moving system, is propagated with the velocity c, if, as we have assumed, this is the case in the stationary system; for we have not as yet furnished the proof that the principle of the constancy of the velocity of light is compatible with the principle of relativity.

At the time $t = \tau = 0$, when the origin of the co-ordinates is common to the two systems, let a spherical wave be emitted therefrom, and be propagated with the velocity c in system K. If (x, y, z) be a point just attained by this wave, then

$$x^2 + y^2 + z^2 = c^2t^2.$$

Transforming this equation with the aid of our equations of transformation we obtain after a simple calculation

$$\xi^2 + \eta^2 + \zeta^2 = c^2\tau^2$$

The wave under consideration is therefore no less a spherical wave with velocity of propagation c when viewed in the moving system. This shows that our two fundamental principles are compatible.*

In the equations of transformation which have been developed there enters an unknown function ϕ of v, which we will now determine.

For this purpose we introduce a third system of co-ordinates K', which relatively to the system k is in a state of parallel translatory motion parallel to the axis of X, such that the origin of co-ordinates of system k moves with velocity $-v$ on the

* The equations of the Lorentz transformation may be more simply deduced directly from the condition that in virtue of these equations the relation $x^2 + y^2 + z^2 = c^2t^2$ shall have as its consequence the second relation $\xi^2 + \eta^2 + \zeta^2 = c^2\tau^2$.

axis of X. At the time $t = 0$ let all three origins coincide, and when $t = x = y = z = 0$ let the time t' of the system K be zero. We call the co-ordinates, measured in the system K', x', y', z', and by a twofold application of our equations of transformation we obtain

$$\begin{aligned} t' &= \phi(-v)\beta(-v)(\tau + v\xi/c^2) &&= \phi(v)\phi(-v)t, \\ x' &= \phi(-v)\beta(-v)(\xi + v\tau) &&= \phi(v)\phi(-v)x, \\ y' &= \phi(-v)\eta &&= \phi(v)\phi(-v)y, \\ z' &= \phi(-v)\zeta &&= \phi(v)\phi(-v)z, \end{aligned}$$

Since the relations between x', y', z' and x, y, z do not contain the time t, the systems K and K' are at rest with respect to one another, and it is clear that the transformation from K to K' must be the identical transformation. Thus

$$\phi(v)\phi(-v) = 1$$

We now inquire into the signification of $\phi(v)$. We give our attention to that part of the axis of Y of system k which lies between $\xi = 0, \eta = 0, \zeta = 0$ and $\xi = 0, \eta = l, \zeta = 0$. This part of the axis of Y is a rod moving perpendicularly to its axis with velocity v relatively to system K. Its ends possess in K the co-ordinates

$$x_1 = vt, y_1 = \frac{l}{\phi(v)}, z_1 = 0$$

and

$$x_2 = vt, y_2 = 0, z_2 = 0$$

The length of the rod measured in K is therefore $l/\phi(v)$; and this gives us the meaning of the function $\phi(v)$. From reasons of symmetry it is now evident that the length of a given rod moving perpendicularly to its axis, measured in the stationary system, must depend only on the velocity and not on the direction and the sense of the motion. The length of the moving rod measured in the stationary system does not change, therefore, if v and $-v$ are interchanged. Hence follows that $l/\phi(v) = l/\phi(-v)$, or

$$\phi(v) = \phi(-v)$$

It follows from this relation and the one previously found that $\phi(v) = 1$, so that the transformation equations which have been found become

$$\begin{aligned} \tau &= \beta\left(t - vx/c^2\right), \\ \xi &= \beta(x - vt), \\ \eta &= y, \\ \zeta &= z \end{aligned}$$

where

$$\beta = 1/\sqrt{(1 - v^2/c^2)}$$

§ 4. Physical Meaning of the Equations Obtained In Respect to Moving Rigid Bodies and Moving Clocks

We envisage a rigid sphere* of radius R, at rest relatively to the moving system *k*, and with its centre at the origin of co-ordinates of *k*. The equation of the surface of this sphere moving relatively to the system K with velocity v is

$$\xi^2 + \eta^2 + \zeta^2 = R^2$$

The equation of this surface expressed in x, y, z at the time $t = 0$ is

$$\frac{x^2}{\left(\sqrt{(1 - v^2/c^2)}\right)^2} + y^2 + z^2 = R^2$$

A rigid body which, measured in a state of rest, has the form of a sphere, therefore has in a state of motion—viewed from the stationary system—the form of an ellipsoid of revolution with the axes

$$R\sqrt{(1 - v^2/c^2)}, R, R$$

Thus, whereas the Y and Z dimensions of the sphere (and therefore of every rigid body of no matter what form) do not appear modified by the motion, the X dimension appears shortened in the ratio $1:\sqrt{(1 - v^2/c^2)}$, i.e. the greater the value of v, the greater the shortening. For $v = c$ all moving objects—viewed from the "stationary" system—shrivel up into plain figures. For velocities greater than that of light our deliberations become meaningless; we shall, however, find in what follows, that the velocity of light in our theory plays the part, physically, of an infinitely great velocity.

It is clear that the same results hold good of bodies at rest in the "stationary" system, viewed from a system in uniform motion.

Further, we imagine one of the clocks which are qualified to mark the time t when at rest relatively to the stationary system, and the time τ when at rest relatively to the moving system, to be located at the origin of the co-ordinates of *k*, and so adjusted that it marks the time τ. What is the rate of this clock, when viewed from the stationary system?

Between the quantities x, t, and τ, which refer to the position of the clock, we have, evidently, $x = vt$ and

* That is, a body possessing spherical form when examined at rest.

$$\tau = \frac{1}{\sqrt{(1 - v^2 / c^2)}} \left(t - vx / c^2\right)$$

Therefore,

$$\tau = t\sqrt{(1 - v^2 / c^2)} = t - (1 - \sqrt{(1 - v^2 / c^2)})t$$

whence it follows that the time marked by the clock (viewed in the stationary system) is slow by $1 - \sqrt{(1-v^2/c^2)}$ seconds per second, or—neglecting magnitudes of fourth and higher order—by $\frac{1}{2}v^2c^2$.

From this there ensues the following peculiar consequence. If at the points A and B of K there are stationary clocks which, viewed in the stationary system, are synchronous; and if the clock at A is moved with the velocity v along the line AB to B, then on its arrival at B the two clocks no longer synchronize, but the clock moved from A to B lags behind the other which has remained at B by $\frac{1}{2}tv^2/c^2$ (up to magnitudes of fourth and higher order), t being the time occupied in the journey from A to B.

It is at once apparent that this result still holds good if the clock moves from A to B in any polygonal line, and also when the points A and B coincide.

If we assume that the result proved for a polygonal line is also valid for a continuously curved line, we arrive at this result: If one of two synchronous clocks at A is moved in a closed curve with constant velocity until it returns to A, the journey lasting t seconds, then by the clock which has remained at rest the travelled clock on its arrival at A will be $\frac{1}{2}tv^2/c^2$ second slow. Thence we conclude that a balance-clock* at the equator must go more slowly, by a very small amount, than a precisely similar clock situated at one of the poles under otherwise identical conditions.

§ 5. THE COMPOSITION OF VELOCITIES

In the system k moving along the axis of X of the system K with velocity v, let a point move in accordance with the equations

$$\xi = w_\xi \tau, \eta = w_\eta \tau, \zeta = 0$$

where w_ξ and w_η denote constants.

Required: the motion of the point relatively to the system K. If with the help of the equations of transformation developed in § 3 we introduce the quantities x, y, z, t into the equations of motion of the point, we obtain

* Not a pendulum-clock, which is physically a system to which the Earth belongs. This case had to be excluded.

$$x = \frac{w_\xi + v}{1 + vw_\xi / c^2} t,$$

$$y = \frac{\sqrt{\left(1 - v^2 / c^2\right)}}{1 + vw_\xi / c^2} w_\eta t,$$

$$z = 0$$

Thus the law of the parallelogram of velocities is valid according to our theory only to a first approximation. We set

$$V^2 = \left(\frac{dx}{dt}\right)^2 + \left(\frac{dy}{dt}\right)^2$$

$$w^2 = w_\xi^{\ 2} + w_\eta^{\ 2},$$

$$a = \tan^{-1} w_y / w_x,$$

a is then to be looked upon as the angle between the velocities v and w. After a simple calculation we obtain

$$V = \frac{\sqrt{\left[\left(v^2 + w^2 + 2vw\cos a\right) - \left(vw \sin a / c^2\right)^2\right]}}{1 + vw \cos a / c^2}$$

It is worthy of remark that v and w enter into the expression for the resultant velocity in a symmetrical manner. If w also has the direction of the axis of X, we get

$$V = \frac{v + w}{1 + vw / c^2}$$

It follows from this equation that from a composition of two velocities which are less than c, there always results a velocity less than c. For if we set $v = c - \kappa$, $w = c - \lambda$, κ and λ being positive and less than c, then

$$V = c \frac{2c - \kappa - \lambda}{2c - \kappa - \lambda + \kappa\lambda/c} < c$$

It follows, further, that the velocity of light c cannot be altered by composition with a velocity less than that of light. For this case we obtain

$$V = \frac{c + w}{1 + w / c} = c$$

We might also have obtained the formula for V, for the case when v and w have the same direction, by compounding two transformations in accordance with § 3. If in addition to the systems K and k figuring in § 3 we introduce still another system of co-ordinates k' moving parallel to k, its initial point moving on the axis of X with the velocity w, we obtain equations between the quantities x, y, z, t and the corresponding quantities of k', which differ from the equations found in § 3 only in that the place of "v" is taken by the quantity

$$\frac{v+w}{1+vw/c^2}$$

from which we see that such parallel transformations—necessarily—form a group.

We have now deduced the requisite laws of the theory of kinematics corresponding to our two principles, and we proceed to show their application to electrodynamics.

II. ELECTRODYNAMICAL PART

§ 6. TRANSFORMATION OF THE MAXWELL-HERTZ EQUATIONS FOR EMPTY SPACE. ON THE NATURE OF THE ELECTROMOTIVE FORCES OCCURRING IN A MAGNETIC FIELD DURING MOTION

Let the Maxwell-Hertz equations for empty space hold good for the stationary system K, so that we have

$$\frac{1}{c}\frac{\partial X}{\partial t}=\frac{\partial N}{\partial y}-\frac{\partial M}{\partial z}, \quad \frac{1}{c}\frac{\partial L}{\partial t}=\frac{\partial Y}{\partial z}-\frac{\partial Z}{\partial y},$$
$$\frac{1}{c}\frac{\partial Y}{\partial t}=\frac{\partial L}{\partial z}-\frac{\partial N}{\partial x}, \quad \frac{1}{c}\frac{\partial M}{\partial t}=\frac{\partial Z}{\partial x}-\frac{\partial X}{\partial z},$$
$$\frac{1}{c}\frac{\partial Z}{\partial t}=\frac{\partial M}{\partial x}-\frac{\partial L}{\partial y}, \quad \frac{1}{c}\frac{\partial N}{\partial t}=\frac{\partial X}{\partial y}-\frac{\partial Y}{\partial x}$$

where (X, Y, Z) denotes the vector of the electric force, and (L, M, N) that of the magnetic force.

If we apply to these equations the transformation developed in § 3, by referring the electromagnetic processes to the system of co-ordinates there introduced, moving with the velocity v, we obtain the equations

$$\frac{1}{c}\frac{\partial X}{\partial \tau} = \frac{\partial}{\partial \eta}\left\{\beta\left(N - \frac{v}{c}Y\right)\right\} - \frac{\partial}{\partial \zeta}\left\{\beta\left(M + \frac{v}{c}Z\right)\right\},$$

$$\frac{1}{c}\frac{\partial}{\partial \tau}\left\{\beta\left(Y - \frac{v}{c}N\right)\right\} = \frac{\partial L}{\partial \xi} - \frac{\partial}{\partial \zeta}\left\{\beta\left(N - \frac{v}{c}Y\right)\right\},$$

$$\frac{1}{c}\frac{\partial}{\partial \tau}\left\{\beta\left(Z + \frac{v}{c}M\right)\right\} = \frac{\partial}{\partial \xi}\left\{\beta\left(M + \frac{v}{c}Z\right)\right\} - \frac{\partial L}{\partial \eta},$$

$$\frac{1}{c}\frac{\partial L}{\partial \tau} = \frac{\partial}{\partial \zeta}\left\{\beta\left(Y - \frac{v}{c}N\right)\right\} - \frac{\partial}{\partial \eta}\left\{\beta\left(Z + \frac{v}{c}M\right)\right\},$$

$$\frac{1}{c}\frac{\partial}{\partial \tau}\left\{\beta\left(M + \frac{v}{c}Z\right)\right\} = \frac{\partial}{\partial \xi}\left\{\beta\left(Z + \frac{v}{c}M\right)\right\} - \frac{\partial X}{\partial \zeta},$$

$$\frac{1}{c}\frac{\partial}{\partial \tau}\left\{\beta\left(N - \frac{v}{c}Y\right)\right\} = \frac{\partial X}{\partial \eta} - \frac{\partial}{\partial \xi}\left\{\beta\left(Y - \frac{v}{c}N\right)\right\},$$

where

$$\beta = 1/\sqrt{(1-v^2/c^2)}$$

Now the principle of relativity requires that if the Maxwell-Hertz equations for empty space hold good in system K, they also hold good in system *k*; that is to say that the vectors of the electric and the magnetic force—(X', Y', Z') and (L', M', N')—of the moving system *k*, which are defined by their ponderomotive effects on electric or magnetic masses respectively, satisfy the following equations:—

$$\frac{1}{c}\frac{\partial X'}{\partial \tau} = \frac{\partial N'}{\partial \eta} - \frac{\partial M'}{\partial \zeta}, \quad \frac{1}{c}\frac{\partial L'}{\partial \tau} = \frac{\partial Y'}{\partial \zeta} - \frac{\partial Z'}{\partial \eta},$$

$$\frac{1}{c}\frac{\partial Y'}{\partial \tau} = \frac{\partial L'}{\partial \zeta} - \frac{\partial N'}{\partial \xi}, \quad \frac{1}{c}\frac{\partial M'}{\partial \tau} = \frac{\partial Z'}{\partial \xi} - \frac{\partial X'}{\partial \zeta},$$

$$\frac{1}{c}\frac{\partial Z'}{\partial \tau} = \frac{\partial M'}{\partial \xi} - \frac{\partial L'}{\partial \eta}, \quad \frac{1}{c}\frac{\partial N'}{\partial \tau} = \frac{\partial X'}{\partial \eta} - \frac{\partial Y'}{\partial \xi}.$$

Evidently the two systems of equations found for system *k* must express exactly the same thing, since both systems of equations are equivalent to the Maxwell-Hertz equations for system K. Since, further, the equations of the two systems agree, with the exception of the symbols for the vectors, it follows that the functions occurring in the systems of equations at corresponding places must agree, with the exception of a factor $\psi(v)$, which is common for all functions of the one system of equations, and is independent of ξ, η, ζ and τ but depends upon v. Thus we have the relations

$$X' = \psi(v)X, \qquad L' = \psi(v)L,$$

$$Y' = \psi(v)\beta\left(Y - \frac{v}{c}N\right), \quad M' = \psi(v)\beta\left(M + \frac{v}{c}Z\right),$$

$$Z' = \psi(v)\beta\left(Z + \frac{v}{c}M\right), \quad N' = \psi(v)\beta\left(N - \frac{v}{c}Y\right).$$

If we now form the reciprocal of this system of equations, firstly by solving the equations just obtained, and secondly by applying the equations to the inverse transformation (from k to K), which is characterized by the velocity $-v$, it follows, when we consider that the two systems of equations thus obtained must be identical, that $\psi(v)\,\psi(-v) = 1$. Further, from reasons of symmetry* $\psi(v) = \psi(-v)$, and therefore

$$\psi(v) = 1,$$

and our equations assume the form

$$X' = X, \qquad L' = L$$

$$Y' = \beta\left(Y - \frac{v}{c}N\right), \quad M' = \beta\left(M + \frac{v}{c}Z\right),$$

$$Z' = \beta\left(Z + \frac{v}{c}M\right), \quad N' = \beta\left(N - \frac{v}{c}Y\right).$$

As to the interpretation of these equations we make the following remarks: Let a point charge of electricity have the magnitude "one" when measured in the stationary system K, i.e. let it when at rest in the stationary system exert a force of one dyne upon an equal quantity of electricity at a distance of one cm. By the principle of relativity this electric charge is also of the magnitude "one" when measured in the moving system. If this quantity of electricity is at rest relatively to the stationary system, then by definition the vector (X, Y, Z) is equal to the force acting upon it. If the quantity of electricity is at rest relatively to the moving system (at least at the relevant instant), then the force acting upon it, measured in the moving system, is equal to the vector (X', Y', Z'). Consequently the first three equations above allow themselves to be clothed in words in the two following ways:

1. If a unit electric point charge is in motion in an electromagnetic field, there acts upon it, in addition to the electric force, an "electromotive force" which, if we neglect the terms multiplied by the second and higher powers of v/c, is equal to the vector-product of the velocity of the charge and the magnetic force, divided by the velocity of light. (Old manner of expression.)

2. If a unit electric point charge is in motion in an electromagnetic field, the force acting upon it is equal to the electric force which is present at the locality of the charge, and which we ascertain by transformation of the field to a system of co-ordinates at rest relatively to the electrical charge. (New manner of expression.)

The analogy holds with "magnetomotive forces." We see that electromotive force plays in the developed theory merely the part of an auxiliary concept, which owes its introduction to the circumstance that electric and magnetic forces do not exist independently of the state of motion of the system of co-ordinates.

* If, for example, $X = Y = Z = L = M = O$, and $N \neq O$, then from reasons of symmetry it is clear that when v changes sign without changing its numerical value, Y' must also change sign without changing its numerical value.

Furthermore it is clear that the asymmetry mentioned in the introduction as arising when we consider the currents produced by the relative motion of a magnet and a conductor, now disappears. Moreover, questions as to the "seat" of electrodynamic electromotive forces (unipolar machines) now have no point.

§ 7. THEORY OF DOPPLER'S PRINCIPLE AND OF ABERRATION

In the system K, very far from the origin of co-ordinates, let there be a source of electrodynamic waves, which in a part of space containing the origin of co-ordinates may be represented to a sufficient degree of approximation by the equations

$$X = X_0 \sin\Phi, \quad L = L_0 \sin\Phi,$$
$$Y = Y_0 \sin\Phi, \quad M = M_0 \sin\Phi,$$
$$Z = Z_0 \sin\Phi, \quad N = N_0 \sin\Phi,$$

where

$$\Phi = \omega\left\{t - \frac{1}{c}(lx + my + nz)\right\}.$$

Here (X_0, Y_0, Z_0) and (L_0, M_0, N_0) are the vectors defining the amplitude of the wave-train, and *l*, *m*, *n* the direction-cosines of the wave-normals. We wish to know the constitution of these waves, when they are examined by an observer at rest in the moving system *k*.

Applying the equations of transformation found in § 6 for electric and magnetic forces, and those found in § 3 for the co-ordinates and the time, we obtain directly

$$X' = X_0 \sin\Phi', \qquad L' = L_0 \sin\Phi',$$
$$Y' = \beta(Y_0 - vN_0/c)\sin\Phi', \quad M' = \beta(M_0 + vZ_0/c)\sin\Phi',$$
$$Z' = \beta(Z_0 + vM_0/c)\sin\Phi', \quad N' = \beta(N_0 - vY_0/c)\sin\Phi',$$
$$\Phi' = \omega'\left\{\tau - \frac{1}{c}(l'\xi + m'\eta + n'\zeta)\right\}$$

where

$$\omega' = \omega\beta(1 - lv/c),$$
$$l' = \frac{l - v/c}{1 - lv/c},$$
$$m' = \frac{m}{\beta(1 - lv/c)},$$
$$n' = \frac{n}{\beta(1 - lv/c)}.$$

From the equation for ω' it follows that if an observer is moving with velocity v relatively to an infinitely distant source of light of frequency ν, in such a way that the connecting line "source—observer" makes the angle ϕ with the velocity of the observer referred to a system of co-ordinates, which is at rest relatively to the source of light, the frequency ν' of the light perceived by the observer is given by the equation

$$\nu' = \nu \frac{1 - \cos\phi \cdot v/c}{\sqrt{\left(1 - v^2/c^2\right)}}.$$

This is Doppler's principle for any velocities whatever. When $\phi = 0$ the equation assumes the perspicuous form

$$\nu' = \nu \sqrt{\frac{1 - v/c}{1 + v/c}}.$$

We see that, in contrast with the customary view, when $v = -c$, $\nu' = \infty$.

If we call the angle between the wave-normal (direction of the ray) in the moving system and the connecting line "source—observer" ϕ', the equation for l assumes the form

$$\cos\phi' = \frac{\cos\phi - v/c}{1 - \cos\phi \cdot v/c}.$$

This equation expresses the law of aberration in its most general form. If $\phi = {}^1/_2\pi$, the equation becomes simply

$$\cos\phi' = -v/c.$$

We still have to find the amplitude of the waves, as it appears in the moving system. If we call the amplitude of the electric or magnetic force A or A' respectively, accordingly as it is measured in the stationary system or in the moving system, we obtain

$$A'^2 = A^2 \frac{\left(1 - \cos\phi \cdot v/c\right)^2}{1 - v^2/c^2}.$$

which equation, if $\phi = 0$, simplifies into

$$A'^2 = A^2 \frac{1 - v/c}{1 + v/c}.$$

It follows from these results that to an observer approaching a source of light with the velocity c, this source of light must appear of infinite intensity.

§ 8. Transformation of the Energy of Light Rays. Theory of the Pressure of Radiation Exerted on Perfect Reflectors

Since $A^2/8\pi$ equals the energy of light per unit of volume, we have to regard $A'^2/8\pi$, by the principle of relativity, as the energy of light in the moving system. Thus A'^2/A^2 would be the ratio of the "measured in motion" to the "measured at rest" energy of a given light complex, if the volume of a light complex were the same, whether measured in K or in *k*. But this is not the case. If *l*, *m*, *n* are the direction-cosines of the wave-normals of the light in the stationary system, no energy passes through the surface elements of a spherical surface moving with the velocity of light:

$$(x - lct)^2 + (y - mct)^2 + (z - nct)^2 = R^2.$$

We may therefore say that this surface permanently encloses the same light complex. We inquire as to the quantity of energy enclosed by this surface, viewed in system *k*, that is, as to the energy of the light complex relatively to the system *k*.

The spherical surface—viewed in the moving system—is an ellipsoidal surface, the equation for which, at the time $\tau = 0$, is

$$(\beta\xi - l\beta\xi v/c)^2 + (\eta - m\beta\xi v/c)^2 + (\zeta - n\beta\xi v/c)^2 = R^2$$

If S is the volume of the sphere, and S' that of this ellipsoid, then by a simple calculation

$$\frac{S'}{S} = \frac{\sqrt{1 - v^2/c^2}}{1 - \cos\phi \cdot v/c}.$$

Thus, if we call the light-energy enclosed by this surface E when it is measured in the stationary system, and E' when measured in the moving system, we obtain

$$\frac{E'}{E} = \frac{A'^2 S'}{A^2 S} = \frac{1 - \cos\phi \cdot v/c}{\sqrt{\left(1 - v^2/c^2\right)}},$$

and this formula, when $\phi = 0$, simplifies into

$$\frac{E'}{E} = \sqrt{\frac{1 - v/c}{1 + v/c}}.$$

It is remarkable that the energy and the frequency of a light complex vary with the state of motion of the observer in accordance with the same law.

Now let the co-ordinate plane $\xi = 0$ be a perfectly reflecting surface, at which the plane waves considered in § 7 are reflected. We seek for the pressure of light exerted on the reflecting surface, and for the direction, frequency, and intensity of the light after reflexion.

Let the incidental light be defined by the quantities A, $\cos\phi$, ν (referred to system K). Viewed from *k* the corresponding quantities are

$$A' = A\frac{1-\cos\phi\cdot v/c}{\sqrt{\left(1-v^2/c^2\right)}},$$

$$\cos\phi' = \frac{\cos\phi - v/c}{1-\cos\phi\cdot v/c},$$

$$v' = v\frac{1-\cos\phi\cdot v/c}{\sqrt{\left(1-v^2/c^2\right)}}.$$

For the reflected light, referring the process to system *k*, we obtain

$$A'' = A'$$
$$\cos\phi'' = -\cos\phi'$$
$$v'' = v'$$

Finally, by transforming back to the stationary system K, we obtain for the reflected light

$$A''' = A''\frac{1+\cos\phi''\cdot v/c}{\sqrt{\left(1-v^2/c^2\right)}} = A\frac{1-2\cos\phi\cdot v/c+v^2/c^2}{1-v^2/c^2},$$

$$\cos\phi''' = \frac{\cos\phi''+v/c}{1+\cos\phi''\cdot v/c} = -\frac{\left(1+v^2/c^2\right)\cos\phi - 2v/c}{1-2\cos\phi\cdot v/c+v^2/c^2}$$

$$v''' = v''\frac{1+\cos\phi''v/c}{\sqrt{\left(1-v^2/c^2\right)}} = v\frac{1-2\cos\phi\cdot v/c+v^2/c^2}{1-v^2/c^2}.$$

The energy (measured in the stationary system) which is incident upon unit area of the mirror in unit time is evidently $A^2(c\cos\phi - v)/8\pi$. The energy leaving the unit of surface of the mirror in the unit of time is $A'''^2(-c\cos\phi''' + v)/8\pi$. The difference of these two expressions is, by the principle of energy, the work done by the pressure of light in the unit of time. If we set down this work as equal to the product Pv, where P is the pressure of light, we obtain

$$P = 2\cdot\frac{A^2}{8\pi}\frac{\left(\cos\phi - v/c\right)^2}{1-v^2/c^2}.$$

In agreement with experiment and with other theories, we obtain to a first approximation

$$P = 2\cdot\frac{A^2}{8\pi}\cos^2\phi.$$

All problems in the optics of moving bodies can be solved by the method here employed. What is essential is that the electric and magnetic force of the light which is influenced by a moving body be transformed into a system of co-ordinates; at rest relatively to the body. By this means all problems in the optics of moving bodies will be reduced to a series of problems in the optics of stationary bodies.

§ 9. TRANSFORMATION OF THE MAXWELL-HERTZ EQUATIONS WHEN CONVECTION-CURRENTS ARE TAKEN INTO ACCOUNT

We start from the equations

$$\frac{1}{c}\left\{\frac{\partial X}{\partial t}+u_x\rho\right\}=\frac{\partial N}{\partial y}-\frac{\partial M}{\partial z},\quad \frac{1}{c}\frac{\partial L}{\partial t}=\frac{\partial Y}{\partial z}-\frac{\partial Z}{\partial y},$$

$$\frac{1}{c}\left\{\frac{\partial Y}{\partial t}+u_y\rho\right\}=\frac{\partial L}{\partial z}-\frac{\partial N}{\partial x},\quad \frac{1}{c}\frac{\partial M}{\partial t}=\frac{\partial Z}{\partial x}-\frac{\partial X}{\partial z},$$

$$\frac{1}{c}\left\{\frac{\partial Z}{\partial t}+u_z\rho\right\}=\frac{\partial M}{\partial x}-\frac{\partial L}{\partial y},\quad \frac{1}{c}\frac{\partial N}{\partial t}=\frac{\partial X}{\partial y}-\frac{\partial Y}{\partial x},$$

where

$$\rho=\frac{\partial X}{\partial x}+\frac{\partial Y}{\partial y}+\frac{\partial Z}{\partial z}$$

denotes 4π times the density of electricity, and (u_x, u_y, u_z) the velocity-vector of the charge. If we imagine the electric charges to be invariably coupled to small rigid bodies (ions, electrons), these equations are the electromagnetic basis of the Lorentzian electrodynamics and optics of moving bodies.

Let these equations be valid in the system K, and transform them, with the assistance of the equations of transformation given in §§ 3 and 6, to the system *k*. We then obtain the equations

$$\frac{1}{c}\left\{\frac{\partial X'}{\partial \tau}+u_\xi\rho'\right\}=\frac{\partial N'}{\partial \eta}-\frac{\partial M'}{\partial \zeta},\quad \frac{1}{c}\frac{\partial L'}{\partial \tau}=\frac{\partial Y'}{\partial \zeta}-\frac{\partial Z'}{\partial \eta},$$

$$\frac{1}{c}\left\{\frac{\partial Y'}{\partial \tau}+u_\eta\rho'\right\}=\frac{\partial L'}{\partial \zeta}-\frac{\partial N'}{\partial \xi},\quad \frac{1}{c}\frac{\partial M'}{\partial \tau}=\frac{\partial Z'}{\partial \xi}-\frac{\partial X'}{\partial \zeta},$$

$$\frac{1}{c}\left\{\frac{\partial Z'}{\partial \tau}+u_\zeta\rho'\right\}=\frac{\partial M'}{\partial \xi}-\frac{\partial L'}{\partial \eta},\quad \frac{1}{c}\frac{\partial N'}{\partial \tau}=\frac{\partial X'}{\partial \eta}-\frac{\partial Y'}{\partial \xi},$$

where

$$u_\xi=\frac{u_x-v}{1-u_xv/c^2}$$

$$u_\eta=\frac{u_y}{\beta\left(1-u_xv/c^2\right)}$$

$$u_\zeta=\frac{u_z}{\beta\left(1-u_xv/c^2\right)},$$

and

$$\rho' = \frac{\partial X'}{\partial \xi} + \frac{\partial Y'}{\partial \eta} + \frac{\partial Z'}{\partial \zeta}$$
$$= \beta\left(1 - u_x v / c^2\right)\rho.$$

Since—as follows from the theorem of addition of velocities (§ 5)—the vector (u_ξ, u_η, u_ζ) is nothing else than the velocity of the electric charge, measured in the system *k*, we have the proof that, on the basis of our kinematical principles, the electrodynamic foundation of Lorentz's theory of the electrodynamics of moving bodies is in agreement with the principle of relativity.

In addition I may briefly remark that the following important law may easily be deduced from the developed equations: If an electrically charged body is in motion anywhere in space without altering its charge when regarded from a system of co-ordinates moving with the body, its charge also remains—when regarded from the "stationary" system K—constant.

§ 10. Dynamics of the Slowly Accelerated Electron

Let there be in motion in an electromagnetic field an electrically charged particle (in the sequel called an "electron"), for the law of motion of which we assume as follows:

If the electron is at rest at a given epoch, the motion of the electron ensues in the next instant of time according to the equations

$$m\frac{d^2x}{dt^2} = \varepsilon X$$
$$m\frac{d^2y}{dt^2} = \varepsilon Y$$
$$m\frac{d^2z}{dt^2} = \varepsilon Z$$

where *x*, *y*, *z* denote the co-ordinates of the electron, and *m* the mass of the electron, as long as its motion is slow.

Now, secondly, let the velocity of the electron at a given epoch be *v*. We seek the law of motion of the electron in the immediately ensuing instants of time.

Without affecting the general character of our considerations, we may and will assume that the electron, at the moment when we give it our attention, is at the origin of the co-ordinates, and moves with the velocity *v* along the axis of X of the system K. It is then clear that at the given moment ($t = 0$) the electron is at rest relatively to a system of co-ordinates which is in parallel motion with velocity *v* along the axis of X.

From the above assumption, in combination with the principle of relativity, it is clear that in the immediately ensuing time (for small values of *t*) the electron, viewed from the system *k*, moves in accordance with the equations

$$m\frac{d^2\xi}{d\tau^2} = \varepsilon X',$$

$$m\frac{d^2\eta}{d\tau^2} = \varepsilon Y',$$

$$m\frac{d^2\zeta}{d\tau^2} = \varepsilon Z',$$

in which the symbols ξ, η, ζ, τ, X', Y', Z' refer to the system *k*. If, further, we decide that when $t = x = y = z = 0$ then $\tau = \xi = \eta = \zeta = 0$, the transformation equations of §§ 3 and 6 hold good, so that we have

$$\xi = \beta(x - vt),\ \eta = y,\ \zeta = z,\ \tau = \beta(t - vx/c^2)$$

$$X' = X,\ Y' = \beta(Y - vN/c),\ Z' = \beta(Z + vM/c).$$

With the help of these equations we transform the above equations of motion from system *k* to system K, and obtain

$$\left.\begin{aligned} \frac{d^2x}{dt^2} &= \frac{\varepsilon}{m\beta^3}X \\ \frac{d^2y}{dt^2} &= \frac{\varepsilon}{m\beta}\left(Y - \frac{v}{c}N\right) \\ \frac{d^2z}{dt^2} &= \frac{\varepsilon}{m\beta}\left(Z + \frac{v}{c}M\right) \end{aligned}\right\} \quad \ldots \text{(A)}$$

Taking the ordinary point of view we now inquire as to the "longitudinal" and the "transverse" mass of the moving electron. We write the equations (A) in the form

$$m\beta^3\frac{d^2x}{dt^2} = \varepsilon X = \varepsilon X',$$

$$m\beta^2\frac{d^2y}{dt^2} = \varepsilon\beta\left(Y - \frac{v}{c}N\right) = \varepsilon Y',$$

$$m\beta^2\frac{d^2z}{dt^2} = \varepsilon\beta\left(Z + \frac{v}{c}M\right) = \varepsilon Z',$$

and remark firstly that $\epsilon X'$, $\epsilon Y'$, $\epsilon Z'$ are the components of the ponderomotive force acting upon the electron, and are so indeed as viewed in a system moving at the moment with the electron, with the same velocity as the electron. (This force might be measured, for example, by a spring balance at rest in the last-mentioned system.) Now if we call this force simply "the force acting upon the electron,"* and maintain the equation—mass $\times$ acceleration = force—and if we also decide that the accelerations are to be measured in the stationary system K, we derive from the above equations

* The definition of force here given is not advantageous, as was first shown by M. Planck. It is more to the point to define force in such a way that the laws of momentum and energy assume the simplest form.

$$\text{Longitudinal mass} = \frac{m}{\left(\sqrt{1-v^2/c^2}\right)^3}.$$

$$\text{Transverse mass} = \frac{m}{1-v^2/c^2}.$$

With a different definition of force and acceleration we should naturally obtain other values for the masses. This shows us that in comparing different theories of the motion of the electron we must proceed very cautiously.

We remark that these results as to the mass are also valid for ponderable material points, because a ponderable material point can be made into an electron (in our sense of the word) by the addition of an electric charge, *no matter how small.*

We will now determine the kinetic energy of the electron. If an electron moves from rest at the origin of co-ordinates of the system K along the axis of X under the action of an electrostatic force X, it is clear that the energy withdrawn from the electrostatic field has the value $\int \varepsilon X dx$. As the electron is to be slowly accelerated, and consequently may not give off any energy in the form of radiation, the energy withdrawn from the electrostatic field must be put down as equal to the energy of motion W of the electron. Bearing in mind that during the whole process of motion which we are considering, the first of the equations (A) applies, we therefore obtain

$$W = \int \varepsilon X dx = m\int_0^v \beta^3 v dv$$

$$= mc^2 \left\{ \frac{1}{\sqrt{1-v^2/c^2}} - 1 \right\}.$$

Thus, when $v = c$, W becomes infinite. Velocities greater than that of light have—as in our previous results—no possibility of existence.

This expression for the kinetic energy must also, by virtue of the argument stated above, apply to ponderable masses as well.

We will now enumerate the properties of the motion of the electron which result from the system of equations (A), and are accessible to experiment.

1. From the second equation of the system (A) it follows that an electric force Y and a magnetic force N have an equally strong deflective action on an electron moving with the velocity v, when $Y = Nv/c$. Thus we see that it is possible by our theory to determine the velocity of the electron from the ratio of the magnetic power of deflexion A_m to the electric power of deflexion A_e, for any velocity, by applying the law

$$\frac{A_m}{A_e} = \frac{v}{c}.$$

This relationship may be tested experimentally, since the velocity of the electron can be directly measured, e.g. by means of rapidly oscillating electric and magnetic fields.

2. From the deduction for the kinetic energy of the electron it follows that between the potential difference, P, traversed and the acquired velocity v of the electron there must be the relationship

$$P = \int X dx = \frac{m}{\varepsilon} c^2 \left\{ \frac{1}{\sqrt{1 - v^2 / c^2}} - 1 \right\}$$

3. We calculate the radius of curvature of the path of the electron when a magnetic force N is present (as the only deflective force), acting perpendicularly to the velocity of the electron. From the second of the equations (A) we obtain

$$-\frac{d^2 y}{dt^2} = \frac{v^2}{R} = \frac{\varepsilon}{m} \frac{v}{c} N \sqrt{1 - \frac{v^2}{c^2}}$$

or

$$R = \frac{mc^2}{\varepsilon} \cdot \frac{v / c}{\sqrt{\left(1 - v^2 / c^2\right)}} \cdot \frac{1}{N}.$$

These three relationships are a complete expression for the laws according to which, by the theory here advanced, the electron must move.

In conclusion I wish to say that in working at the problem here dealt with I have had the loyal assistance of my friend and colleague M. Besso, and that I am indebted to him for several valuable suggestions.

DOES THE INERTIA OF A BODY DEPEND UPON ITS ENERGY-CONTENT?

Translated from "Ist die Trägheit eins Körpers von seinem Energiegehalt abhängig?" Annalen der Physik, 17, 1905.

The results of the previous investigation lead to a very interesting conclusion, which is here to be deduced.

I based that investigation on the Maxwell-Hertz equations for empty space, together with the Maxwellian expression for the electromagnetic energy of space, and in addition the principle that:

The laws by which the states of physical system alter are independent of the alternative, to which of two systems of co-ordinates, in uniform motion of parallel translation relatively to each other, these alterations of state are referred (principle of relativity).

With these principles* as my basis I deduced *inter alia* the following result (§ 8):

Let a system of plane waves of light, referred to the system of co-ordinates (x, y, z), possess the energy l; let the direction of the ray (the wave-normal) make an angle ϕ with the axis of x of the system. If we introduce a new system of co-ordinates (ξ, η, ζ) moving in uniform parallel translation with respect to the system (x, y, z), and having its origin of co-ordinates in motion along the axis of x with the velocity v, then this quantity of light—measured in the system (ξ, η, ζ)—possesses the energy

$$l^* = l \frac{1 - \frac{v}{c}\cos\phi}{\sqrt{1 - v^2/c^2}}$$

where c denotes the velocity of light. We shall make use of this result in what follows.

Let there be a stationary body in the system (x, y, z), and let its energy—referred to the system (x, y, z)—be E_0. Let the energy of the body relative to the system (ξ, η, ζ) moving as above with the velocity v, be H_0.

Let this body send out, in a direction making an angle ϕ with the axis of x, plane waves of light, of energy $^1/_2 L$ measured relatively to (x, y, z), and simultaneously an equal quantity of light in the opposite direction. Meanwhile the body remains at rest with respect to the system (x, y, z). The principle of energy must apply to this process, and in fact (by the principle of relativity) with respect to both systems of co-ordinates. If we call the energy of the body after the emission of light E_1 or H_1 respectively, measured relatively to the system (x, y, z) or (ξ, η, ζ) respectively, then by employing the relation given above we obtain

$$E_0 = E_1 + \frac{1}{2}L + \frac{1}{2}L,$$

$$\begin{aligned} H_0 &= H_1 + \frac{1}{2}L\frac{1 - \frac{v}{c}\cos\phi}{\sqrt{1 - v^2/c^2}} + \frac{1}{2}L\frac{1 + \frac{v}{c}\cos\phi}{\sqrt{1 - v^2/c^2}} \\ &= H_1 + \frac{L}{\sqrt{1 - v^2/c^2}} \end{aligned}$$

By subtraction we obtain from these equations

*The principle of the constancy of the velocity of light is of course contained in Maxwell's equations.

$$H_0 - E_0 - (H_1 - E_1) = L\left\{\frac{1}{\sqrt{1 - v^2/c^2}} - 1\right\}.$$

The two differences of the form H – E occurring in this expression have simple physical significations. H and E are energy values of the same body referred to two systems of co-ordinates which are in motion relatively to each other, the body being at rest in one of the two systems (system (x, y, z)). Thus it is clear that the difference H – E can differ from the kinetic energy K of the body, with respect to the other system (ξ, η, ζ), only by an additive constant C, which depends on the choice of the arbitrary additive constants of the energies H and E. Thus we may place

$$H_0 - E_0 = K_0 + C,$$

$$H_1 - E_1 = K_1 + C,$$

since C does not change during the emission of light. So we have

$$K_0 - K_1 = L\left\{\frac{1}{\sqrt{1 - v^2/c^2}} - 1\right\}.$$

The kinetic energy of the body with respect to (ξ, η, ζ) diminishes as a result of the emission of light, and the amount of diminution is independent of the properties of the body. Moreover, the difference K_0 - K_1, like the kinetic energy of the electron (§ 10), depends on the velocity.

Neglecting magnitudes of fourth and higher orders we may place

$$K_0 - K_1 = \frac{1}{2}\frac{L}{c^2}v^2 \quad .$$

From this equation it directly follows that:

If a body gives off the energy L in the form of radiation, its mass diminishes by L/c^2. The fact that the energy withdrawn from the body becomes energy of radiation evidently makes no difference, so that we are led to the more general conclusion that:

The mass of a body is a measure of its energy-content; if the energy changes by L, the mass changes in the same sense by $L/9 \times 10^{20}$, the energy being measured in ergs, and the mass in grammes.

It is not impossible that with bodies whose energy-content is variable to a high degree (e.g. with radium salts) the theory may be successfully put to the test.

If the theory corresponds to the facts, radiation conveys inertia between the emitting and absorbing bodies.

ON THE INFLUENCE OF GRAVITATION ON THE PROPAGATION OF LIGHT

Translated from "Über den Einfluss der Schwerkraft auf die Ausbreitung des Lichtes," Annalen der Physik, 35, 1911.

In a memoir published four years ago* I tried to answer the question whether the propagation of light is influenced by gravitation. I return to this theme, because my previous presentation of the subject does not satisfy me, and for a stronger reason, because I now see that one of the most important consequences of my former treatment is capable of being tested experimentally. For it follows from the theory here to be brought forward, that rays of light, passing close to the sun, are deflected by its gravitational field, so that the angular distance between the sun and a fixed star appearing near to it is apparently increased by nearly a second of arc.

In the course of these reflexions further results are yielded which relate to gravitation. But as the exposition of the entire group of considerations would be rather difficult to follow, only a few quite elementary reflexions will be given in the following pages, from which the reader will readily be able to inform himself as to the suppositions of the theory and its line of thought. The relations here deduced, even if the theoretical foundation is sound, are valid only to a first approximation.

§ 1. A HYPOTHESIS AS TO THE PHYSICAL NATURE OF THE GRAVITATIONAL FIELD

In a homogeneous gravitational field (acceleration of gravity γ) let there be a stationary system of co-ordinates K, orientated so that the lines of force of the gravitational field run in the negative direction of the axis of z. In a space free of gravitational fields let there be a second system of co-ordinates K', moving with uniform acceleration (γ) in the positive direction of its axis of z. To avoid unnecessary complications, let us for the present disregard the theory of relativity, and regard both systems from the customary point of view of kinematics, and the movements occurring in them from that of ordinary mechanics.

Relatively to K, as well as relatively to K', material points which are not subjected to the action of other material points, move in keeping with the equations

* A. Einstein, Jahrbuch für Radioakt. und Elektronik, 4, 1907

$$\frac{d^2x}{dt^2} = 0, \frac{d^2y}{dt^2} = 0, \frac{d^2z}{dt^2} = -\gamma.$$

For the accelerated system K' this follows directly from Galileo's principle, but for the system K, at rest in a homogeneous gravitational field, from the experience that all bodies in such a field are equally and uniformly accelerated. This experience, of the equal falling of all bodies in the gravitational field, is one of the most universal which the observation of nature has yielded; but in spite of that the law has not found any place in the foundations of our edifice of the physical universe.

But we arrive at a very satisfactory interpretation of this law of experience, if we assume that the systems K and K' are physically exactly equivalent, that is, if we assume that we may just as well regard the system K as being in a space free from gravitational fields, if we then regard K as uniformly accelerated. This assumption of exact physical equivalence makes it impossible for us to speak of the absolute acceleration of the system of reference, just as the usual theory of relativity forbids us to talk of the absolute velocity of a system;* and it makes the equal falling of all bodies in a gravitational field seem a matter of course.

As long as we restrict ourselves to purely mechanical processes in the realm where Newton's mechanics holds sway, we are certain of the equivalence of the systems K and K'. But this view of ours will not have any deeper significance unless the systems K and K' are equivalent with respect to all physical processes, that is, unless the laws of nature with respect to K are in entire agreement with those with respect to K'. By assuming this to be so, we arrive at a principle which, if it is really true, has great heuristic importance. For by theoretical consideration of processes which take place relatively to a system of reference with uniform acceleration, we obtain information as to the career of processes in a homogeneous gravitational field. We shall now show, first of all, from the standpoint of the ordinary theory of relativity, what degree of probability is inherent in our hypothesis.

§ 2. ON THE GRAVITATION OF ENERGY

One result yielded by the theory of relativity is that the inertia mass of a body increases with the energy it contains; if the increase of energy amounts to E, the increase in inertia mass is equal to E/c^2, when c denotes the velocity of light. Now is there an increase of gravitating mass corresponding to this increase of inertia mass? If not, then a body would fall in the same gravitational field with varying acceleration according to the energy it contained. That highly satisfactory result of the theory of

* Of course we cannot replace any arbitrary gravitational field by a state of motion of the system without a gravitational field, any more than, by a transformation of relativity, we can transform all points of a medium in any kind of motion to rest.

relativity by which the law of the conservation of mass is merged in the law of conservation of energy could not be maintained, because it would compel us to abandon the law of the conservation of mass in its old form for inertia mass, and maintain it for gravitating mass.

But this must be regarded as very improbable. On the other hand, the usual theory of relativity does not provide us with any argument from which to infer that the weight of a body depends on the energy contained in it. But we shall show that our hypothesis of the equivalence of the systems K and K' gives us gravitation of energy as a necessary consequence.

Let the two material systems S_1 and S_2, provided with instruments of measurement, be situated on the *z*-axis of K at the distance *h* from each other,* so that the gravitation potential in S_2 is greater than that in S_1 by γh. Let a definite quantity of energy E be emitted from S_2 towards S_1. Let the quantities of energy in S_1 and S_2 be measured by contrivances which—brought to one place in the system *z* and there compared—shall be perfectly alike. As to the process of this conveyance of energy by radiation we can make no *a priori* assertion, because we do not know the influence of the gravitational field on the radiation and the measuring instruments in S_1 and S_2.

But by our postulate of the equivalence of K and K' we are able, in place of the system K in a homogeneous gravitational field, to set the gravitation-free system K', which moves with uniform acceleration in the direction of positive *z*, and with the *z*-axis of which the material systems S_1 and S_2 are rigidly connected.

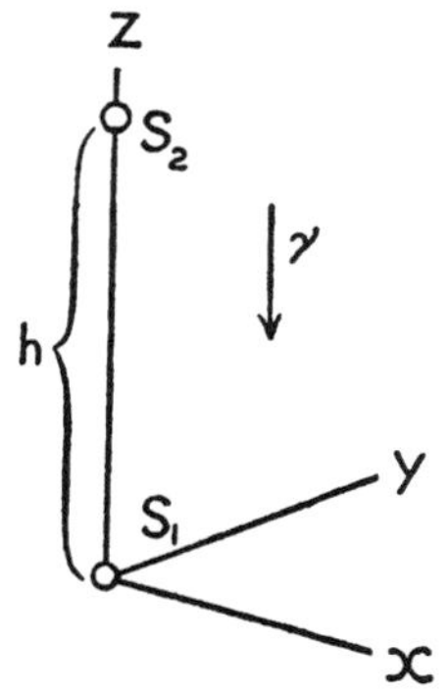

Fig. 5

We judge of the process of the transference of energy by radiation from S_2 to S_1 from a system K_0, which is to be free from acceleration. At the moment when the radiation energy E_2 is emitted from S_2 toward S_1, let the velocity of K' relatively to K_0 be zero. The radiation will arrive at S_1 when the time h/c has elapsed (to a first approximation). But at this moment the velocity of S_1 relatively to K_0 is $\gamma h/c = v$. Therefore by the ordinary theory of relativity the radiation arriving at S_1 does not possess the energy E_2, but a greater energy E_1, which is related to E_2, to a first approximation by the equation†

$$E_1 = E_2\left(1 + \frac{v}{c}\right) = E_2\left(1 + \gamma\frac{h}{c^2}\right) \qquad (1)$$

By our assumption exactly the same relation holds if the same process takes place in the system K, which is not accelerated, but is provided with a gravitational field. In this case we may replace γh by the potential Φ of the gravitation vector in S_2, if the arbitrary constant of Φ in S_1 is equated to zero. We then have the equation

* The dimensions of S1 and S2, are regarded as infinitely small in comparison with h.

† See above.

$$E_1 = E_2 + \frac{E_2}{c^2}\Phi \qquad (1a)$$

This equation expresses the law of energy for the process under observation. The energy E_1 arriving at S_1 is greater than the energy E_2, measured by the same means, which was emitted in S_2, the excess being the potential energy of the mass E_2/c^2 in the gravitational field. It thus proves that for the fulfilment of the principle of energy we have to ascribe to the energy E, before its emission in S_2, a potential energy due to gravity, which corresponds to the gravitational mass E/c^2. Our assumption of the equivalence of K and K' thus removes the difficulty mentioned at the beginning of this paragraph which is left unsolved by the ordinary theory of relativity.

The meaning of this result is shown particularly clearly if we consider the following cycle of operations:

1. The energy E, as measured in S_2, is emitted in the form of radiation in S_2 towards S_1, where, by the result just obtained, the energy $E(1 + \gamma h/c^2)$, as measured in S_1, is absorbed.

2. A body W of mass M is lowered from S_2 to S_1, work $M\gamma h$ being done in the process.

3. The energy E is transferred from S_1 to the body W while W is in S_1. Let the gravitational mass M be thereby changed so that it acquires the value M'.

4. Let W be again raised to S_2, work $M'\gamma h$ being done in the process.

5. Let E be transferred from W back to S_2.

The effect of this cycle is simply that S_1 has undergone the increase of energy $E\gamma h/c^2$, and that the quantity of energy $M'\gamma k - M\gamma h$ has been conveyed to the system in the form of mechanical work. By the principle of energy, we must therefore have

$$E\gamma \frac{h}{c^2} = M'\gamma h - M\gamma h,$$

or

$$M' - M = E/c^2 \qquad (1b)$$

The increase in gravitational mass is thus equal to E/c^2, and therefore equal to the increase in inertia mass as given by the theory of relativity.

The result emerges still more directly from the equivalence of the systems K and K', according to which the gravitational mass in respect of K is exactly equal to the inertia mass in respect of K'; energy must therefore possess a gravitational mass which is equal to its inertia mass. If a mass M_0 be suspended on a spring balance in the system K', the balance will indicate the apparent weight $M_0\gamma$ on account of the inertia of M_0. If the quantity of energy E be transferred to M_0, the spring balance, by the law of the inertia of energy, will indicate $(M_0 + E/c^2)\gamma$. By reason of our fundamental assumption exactly the same thing must occur when the experiment is repeated in the system K, that is, in the gravitational field.

3. TIME AND THE VELOCITY OF LIGHT IN THE GRAVITATIONAL FIELD

If the radiation emitted in the uniformly accelerated system K' in S_2 toward S_1 had the frequency ν_2 relatively to the clock in S_2, then, relatively to S_1, at its arrival in S_1 it no longer has the frequency ν_2, relatively to an identical clock in S_1, but a greater frequency ν_1, such that to a first approximation

$$\nu_1 = \nu_2\left(1 + \gamma\frac{h}{c^2}\right). \tag{2}$$

For if we again introduce the unaccelerated system of reference K_0, relatively to which, at the time of the emission of light, K' has no velocity, then S_1, at the time of arrival of the radiation at S_1, has, relatively to K_0, the velocity $\gamma h/c$, from which, by Doppler's principle, the relation as given results immediately.

In agreement with our assumption of the equivalence of the systems K' and K, this equation also holds for the stationary system of co-ordinates K, provided with a uniform gravitational field, if in it the transference by radiation takes place as described. It follows, then, that a ray of light emitted in S_2 with a definite gravitational potential, and possessing at its emission the frequency ν_2—compared with a clock in S_2—will, at its arrival in S_1, possess a different frequency ν_1—measured by an identical clock in S_1. For γh we substitute the gravitational potential Φ of S_2—that of S_1 being taken as zero—and assume that the relation which we have deduced for the homogeneous gravitational field also holds for other forms of field. Then

$$\nu_1 = \nu_2\left(1 + \frac{\Phi}{c^2}\right) \tag{2a}$$

This result (which by our deduction is valid to a first approximation) permits, in the first place, of the following application. Let ν_0 be the vibration-number of an elementary light-generator, measured by a delicate clock at the same place. Let us imagine them both at a place on the surface of the Sun (where our S_2 is located). Of the light there emitted, a portion reaches the Earth (S_1), where we measure the frequency of the arriving light with a clock U in all respects resembling the one just mentioned. Then by (2a),

$$\nu = \nu_0\left(1 + \frac{\Phi}{c^2}\right)$$

where Φ is the (negative) difference of gravitational potential between the surface of the Sun and the Earth. Thus according to our view the spectral lines of sunlight, as compared with the corresponding spectral lines of terrestrial sources of light, must be somewhat displaced toward the red, in fact by the relative amount

$$\frac{\nu_0 - \nu}{\nu_0} = -\frac{\Phi}{c^2} = 2.10^{-6}$$

If the conditions under which the solar bands arise were exactly known, this shifting would be susceptible of measurement. But as other influences (pressure, temperature) affect the position of the centres of the spectral lines, it is difficult to discover whether the inferred influence of the gravitational potential really exists.*

On a superficial consideration equation (2), or (2a), respectively, seems to assert an absurdity. If there is constant transmission of light from S_2 to S_1, how can any other number of periods per second arrive in S_1 than is emitted in S_2? But the answer is simple. We cannot regard ν_2 or respectively ν_1 simply as frequencies (as the number of periods per second) since we have not yet determined the time in system K. What ν_2 denotes is the number of periods with reference to the time-unit of the clock U in S_2, while ν_1 denotes the number of periods per second with reference to the identical clock in S_1. Nothing compels us to assume that the clocks U in different gravitation potentials must be regarded as going at the same rate. On the contrary, we must certainly define the time in K in such a way that the number of wave crests and troughs between S_2 and S_1 is independent of the absolute value of time; for the process under observation is by nature a stationary one. If we did not satisfy this condition, we should arrive at a definition of time by the application of which time would merge explicitly into the laws of nature, and this would certainly be unnatural and unpractical. Therefore the two clocks in S_1 and S_2 do not both give the "time" correctly. If we measure time in S_1 with the clock U, then we must measure time in S_2 with a clock which goes $1 + \Phi/c^2$ times more slowly than the clock U when compared with U at one and the same place. For when measured by such a clock the frequency of the ray of light which is considered above is at its emission in S_2

$$\nu_2\left(1 + \frac{\Phi}{c^2}\right)$$

and is therefore, by (2a), equal to the frequency ν_1 of the same ray of light on its arrival in S_1.

This has a consequence which is of fundamental importance for our theory. For if we measure the velocity of light at different places in the accelerated, gravitation-free system K', employing clocks U of identical constitution, we obtain the same magnitude at all these places. The same holds good, by our fundamental assumption, for the system K as well. But from what has just been said we must use clocks of unlike

* L. F. Jewell (Journ. de Phys., 6, 1897, p. 84) and particularly Ch. Fabry and H. Boisson (Comptes rendus, 148, 1909, pp. 688–690) have actually found such displacements of fine spectral lines toward the red end of the spectrum, of the order of magnitude here calculated, but have ascribed them to an effect of pressure in the absorbing layer.

constitution, for measuring time at places with differing gravitation potential. For measuring time at a place which, relatively to the origin of the co-ordinates, has the gravitation potential Φ, we must employ a clock which—when removed to the origin of co-ordinates—goes $(1 + \Phi/c^2)$ times more slowly than the clock used for measuring time at the origin of co-ordinates. If we call the velocity of light at the origin of co-ordinates c_0, then the velocity of light c at a place with the gravitation potential Φ will be given by the relation

$$c = c_0\left(1 + \frac{\Phi}{c^2}\right) \qquad (3)$$

The principle of the constancy of the velocity of light holds good according to this theory in a different form from that which usually underlies the ordinary theory of relativity.

4. Bending of Light-Rays in the Gravitational Field

From the proposition which has just been proved, that the velocity of light in the gravitational field is a function of the place, we may easily infer, by means of Huyghens's principle, that light-rays propagated across a gravitational field undergo deflexion. For let E be a wave front of a plane light-wave at the time t, and let P_1 and P_2 be two points in that plane at

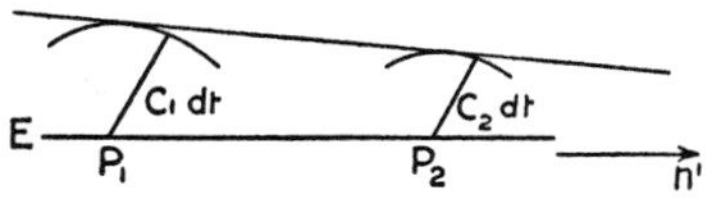

Fig. 6

unit distance from each other. P_1 and P_2 lie in the plane of the paper, which is chosen so that the differential coefficient of Φ, taken in the direction of the normal to the plane, vanishes, and therefore also that of c. We obtain the corresponding wave front at time $t + dt$, or, rather, its line of section with the plane of the paper, by describing circles round the points P_1 and P_2 with radii $c_1 dt$ and $c_2 dt$ respectively, where c_1 and c_2 denote the velocity of light at the points P_1 and P_2 respectively, and by drawing the tangent to these circles. The angle through which the light-ray is deflected in the path cdt is therefore

$$(c_1 - c_2)dt = -\frac{\partial c}{\partial n'}dt,$$

if we calculate the angle positively when the ray is bent toward the side of increasing n'. The angle of deflexion per unit of path of the light-ray is thus

$$-\frac{1}{c}\frac{\partial c}{\partial n'} \text{ or by (3) } -\frac{1}{c^2}\frac{\partial \Phi}{\partial n'}$$

Finally, we obtain for the deflexion which a light-ray experiences toward the side n' on any path (s) the expression

$$a = -\frac{1}{c^2}\int\frac{\partial \Phi}{\partial n'}ds \tag{4}$$

We might have obtained the same result by directly considering the propagation of a ray of light in the uniformly accelerated system K', and transferring the result to the system K, and thence to the case of a gravitational field of any form.

By equation (4) a ray of light passing along by a heavenly body suffers a deflexion to the side of the diminishing gravitational potential, that is, on the side directed toward the heavenly body, of the magnitude

$$a = -\frac{1}{c^2}\int_{\theta=-\frac{1}{2}\pi}^{\theta=\frac{1}{2}\pi}\frac{kM}{r^2}\cos\theta ds = 2\frac{kM}{c^2\Delta}$$

where k denotes the constant of gravitation, M the mass of the heavenly body, Δ the distance of the ray from the centre of the body. A ray of light going past the Sun would accordingly undergo deflexion to the amount of $4 \cdot 10^{-6}$ = .83 seconds of arc. The angular distance of the star from the centre of the Sun appears to be increased by this amount. As the fixed stars in the parts of the sky near the Sun are visible during total eclipses of the Sun, this consequence of the theory may be compared with experience. With the planet Jupiter the displacement to be expected reaches to about $^1/_{100}$ of the amount given. It would be a most desirable thing if astronomers would take up the question here raised. For apart from any theory there is the question whether it is possible with the equipment at present available to detect an influence of gravitational fields on the propagation of light.

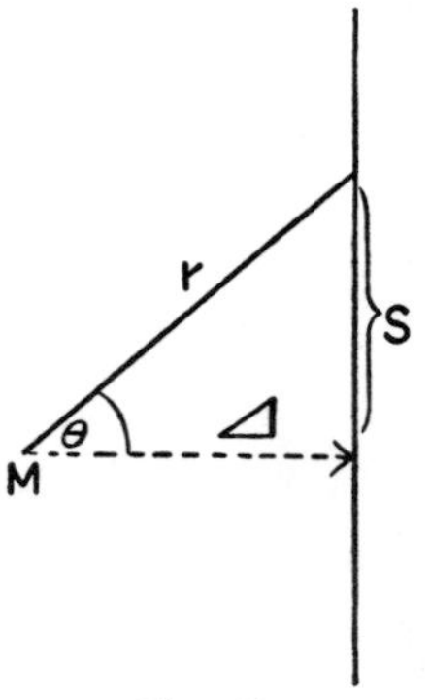

Fig 7

THE FOUNDATION OF THE GENERAL THEORY OF RELATIVITY

Translated from "Die Grundlage der allgemeinen Relativitätstheorie," Annalen der Physik, 49, 1916.

A. FUNDAMENTAL CONSIDERATIONS ON THE POSTULATE OF RELATIVITY

§ 1. OBSERVATIONS ON THE SPECIAL THEORY OF RELATIVITY

The special theory of relativity is based on the following postulate, which is also satisfied by the mechanics of Galileo and Newton.

If a system of co-ordinates K is chosen so that, in relation to it, physical laws hold good in their simplest form, the *same* laws also hold good in relation to any other system of co-ordinates K' moving in uniform translation relatively to K. This postulate we call the "special principle of relativity." The word "special" is meant to intimate that the principle is restricted to the case when K' has a motion of uniform translation relatively to K, but that the equivalence of K' and K does not extend to the case of non-uniform motion of K' relatively to K.

Thus the special theory of relativity does not depart from classical mechanics through the postulate of relativity, but through the postulate of the constancy of the velocity of light *in vacuo*, from which, in combination with the special principle of relativity, there follow, in the well-known way, the relativity of simultaneity, the Lorentzian transformation, and the related laws for the behaviour of moving bodies and clocks.

The modification to which the special theory of relativity has subjected the theory of space and time is indeed far-reaching, but one important point has remained unaffected. For the laws of geometry, even according to the special theory of relativity, are to be interpreted directly as laws relating to the possible relative positions of solid bodies at rest; and, in a more general way, the laws of kinematics are to be interpreted as laws which describe the relations of measuring bodies and clocks. To two selected material points of a stationary rigid body there always corresponds a distance of quite definite length, which is independent of the locality and orientation of the body, and is also independent of the time. To two selected positions of the hands of a clock at rest relatively to the privileged system of reference there always corresponds an interval of time of a definite length, which is independent of place and time. We shall soon see that the general theory of relativity cannot adhere to this simple physical interpretation of space and time.

§ 2. THE NEED FOR AN EXTENSION OF THE POSTULATE OF RELATIVITY

In classical mechanics, and no less in the special theory of relativity, there is an inherent epistemological defect which was, perhaps for the first time, clearly pointed out by Ernst Mach. We will elucidate it by the following example: Two fluid bodies of the same size and nature hover freely in space at so great a distance from each other and from all other masses that only those gravitational forces need be taken into account which arise from the interaction of different parts of the same body. Let the distance between the two bodies be invariable, and in neither of the bodies let there be any relative movements of the parts with respect to one another. But let either mass, as judged by an observer at rest relatively to the other mass, rotate with constant angular velocity about the line joining the masses. This is a verifiable relative motion of the two bodies. Now let us imagine that each of the bodies has been surveyed by means of measuring instruments at rest relatively to itself, and let the surface of S_1 prove to be a sphere, and that of S_2 an ellipsoid of revolution. Thereupon we put the question—What is the reason for this difference in the two bodies? No answer can be admitted as epistemologically satisfactory,* unless the reason given is an *observable fact of experience.* The law of causality has not the significance of a statement as to the world of experience, except when *observable facts* ultimately appear as causes and effects.

Newtonian mechanics does not give a satisfactory answer to this question. It pronounces as follows: The laws of mechanics apply to the space R_1, in respect to which the body S_1 is at rest, but not to the space R_2 in respect to which the body S_2 is at rest. But the privileged space R_1 of Galileo, thus introduced, is a merely *factitious* cause, and not a thing that can be observed. It is therefore clear that Newton's mechanics does not really satisfy the requirement of causality in the case under consideration, but only apparently does so, since it makes the factitious cause R_1 responsible for the observable difference in the bodies S_1 and S_2.

The only satisfactory answer must be that the physical system consisting of S_1 and S_2 reveals within itself no imaginable cause to which the differing behaviour of S_1 and S_2 can be referred. The cause must therefore lie *outside* this system. We have to take it that the general laws of motion, which in particular determine the shapes of S_1 and S_2, must be such that the mechanical behaviour of S_1 and S_2 is partly conditioned, in quite essential respects, by distant masses which we have not included in the system under consideration. These distant masses and their motions relative to S_1 and S_2 must then be regarded as the seat of the causes (which must be susceptible to observation) of the different behaviour of our two bodies S_1 and S_2. They take over the rôle of the factitious cause R_1. Of all imaginable spaces R_1, R_2, etc., in any kind of motion relatively to one another, there is none which we may look upon as privileged *a priori* without reviving the above-mentioned epistemological objection. *The laws of physics must be of such a nature that they apply to systems of reference in any kind of motion.* Along this road we arrive at an extension of the postulate of relativity.

* Of course an answer may be satisfactory from the point of view of epistemology, and yet be unsound physically, if it is in conflict with other experiences.

In addition to this weighty argument from the theory of knowledge, there is a well-known physical fact which favours an extension of the theory of relativity. Let K be a Galilean system of reference, i.e. a system relatively to which (at least in the four-dimensional region under consideration) a mass, sufficiently distant from other masses, is moving with uniform motion in a straight line. Let K' be a second system of reference which is moving relatively to K in *uniformly accelerated* translation. Then, relatively to K', a mass sufficiently distant from other masses would have an accelerated motion such that its acceleration and direction of acceleration are independent of the material composition and physical state of the mass.

Does this permit an observer at rest relatively to K' to infer that he is on a "really" accelerated system of reference? The answer is in the negative; for the above-mentioned relation of freely movable masses to K' may be interpreted equally well in the following way. The system of reference K' is unaccelerated, but the space-time territory in question is under the sway of a gravitational field, which generates the accelerated motion of the bodies relatively to K'.

This view is made possible for us by the teaching of experience as to the existence of a field of force, namely, the gravitational field, which possesses the remarkable property of imparting the same acceleration to all bodies.* The mechanical behaviour of bodies relatively to K' is the same as presents itself to experience in the case of systems which we are wont to regard as "stationary" or as "privileged." Therefore, from the physical standpoint, the assumption readily suggests itself that the systems K and K' may both with equal right be looked upon as "stationary," that is to say, they have an equal title as systems of reference for the physical description of phenomena.

It will be seen from these reflexions that in pursuing the general theory of relativity we shall be led to a theory of gravitation, since we are able to "produce" a gravitational field merely by changing the system of co-ordinates. It will also be obvious that the principle of the constancy of the velocity of light *in vacuo* must be modified, since we easily recognize that the path of a ray of light with respect to K' must in general be curvilinear, if with respect to K light is propagated in a straight line with a definite constant velocity.

§ 3. THE SPACE-TIME CONTINUUM. REQUIREMENT OF GENERAL CO-VARIANCE FOR THE EQUATIONS EXPRESSING GENERAL LAWS OF NATURE

In classical mechanics, as well as in the special theory of relativity, the co-ordinates of space and time have a direct physical meaning. To say that a point-event has the X_1 co-ordinate x_1 means that the projection of the point-event on the axis of X_1, determined by rigid rods and in accordance with the rules of Euclidean geometry, is obtained by measuring off a given rod (the unit of length) x_1 times from the origin of

* Eötvös has proved experimentally that the gravitational field has this property in great accuracy.

co-ordinates along the axis of X_1. To say that a point-event has the X_4 co-ordinates $x_4 = t$, means that a standard clock, made to measure time in a definite unit period, and which is stationary relatively to the system of co-ordinates and practically coincident in space with the point-event,* will have measured off $x_4 = t$ periods at the occurrence of the event.

This view of space and time has always been in the minds of physicists, even if, as a rule, they have been unconscious of it. This is clear from the part which these concepts play in physical measurements; it must also have underlain the reader's reflexions on the preceding paragraph (§ 2) for him to connect any meaning with what he there read. But we shall now show that we must put it aside and replace it by a more general view, in order to be able to carry through the postulate of general relativity, if the special theory of relativity applies to the special case of the absence of a gravitational field.

In a space which is free of gravitational fields we introduce a Galilean system of reference K (x, y, z, t), and also a system of co-ordinates K' (x', y', z', t') in uniform rotation relatively to K. Let the origins of both systems, as well as their axes of Z, permanently coincide. We shall show that for a space-time measurement in the system K' the above definition of the physical meaning of lengths and times cannot be maintained. For reasons of symmetry it is clear that a circle around the origin in the X, Y plane of K may at the same time be regarded as a circle in the X', Y' plane of K'. We suppose that the circumference and diameter of this circle have been measured with a unit measure infinitely small compared with the radius, and that we have the quotient of the two results. If this experiment were performed with a measuring-rod at rest relatively to the Galilean system K, the quotient would be π. With a measuring-rod at rest relatively to K', the quotient would be greater than π. This is readily understood if we envisage the whole process of measuring from the "stationary" system K, and take into consideration that the measuring-rod applied to the periphery undergoes a Lorentzian contraction, while the one applied along the radius does not. Hence Euclidean geometry does not apply to K'. The notion of co-ordinates defined above, which presupposes the validity of Euclidean geometry, therefore breaks down in relation to the system K'. So, too, we are unable to introduce a time corresponding to physical requirements in K', indicated by clocks at rest, relatively to K'. To convince ourselves of this impossibility, let us imagine two clocks of identical constitution placed, one at the origin of co-ordinates, and the other at the circumference of the circle, and both envisaged from the "stationary" system K. By a familiar result of the special theory of relativity, the clock at the circumference—judged from K—goes more slowly than the other, because the former is in motion and the latter at rest. An observer at the common origin of co-ordinates, capable of observing the clock at the circumference by means of light, would therefore

* We assume the possibility of verifying "simultaneity" for events immediately proximate in space, or—to speak more precisely—for immediate proximity or coincidence in space-time, without giving a definition of this fundamental concept.

see it lagging behind the clock beside him. As he will not make up his mind to let the velocity of light along the path in question depend explicitly on the time, he will interpret his observations as showing that the clock at the circumference "really" goes more slowly than the clock at the origin. So he will be obliged to define time in such a way that the rate of a clock depends upon where the clock may be.

We therefore reach this result:—In the general theory of relativity, space and time cannot be defined in such a way that differences of the spatial co-ordinates can be directly measured by the unit measuring-rod, or differences in the time co-ordinate by a standard clock.

The method hitherto employed for laying co-ordinates into the space-time continuum in a definite manner thus breaks down, and there seems to be no other way which would allow us to adapt systems of co-ordinates to the four-dimensional universe so that we might expect from their application a particularly simple formulation of the laws of nature. So there is nothing for it but to regard all imaginable systems of co-ordinates, on principle, as equally suitable for the description of nature. This comes to requiring that:—

The general laws of nature are to be expressed by equations which hold good for all systems of co-ordinates, that is, are co-variant with respect to any substitutions whatever (generally co-variant).

It is clear that a physical theory which satisfies this postulate will also be suitable for the general postulate of relativity. For the sum of *all* substitutions in any case includes those which correspond to all relative motions of three-dimensional systems of co-ordinates. That this requirement of general co-variance, which takes away from space and time the last remnant of physical objectivity, is a natural one, will be seen from the following reflexion. All our space-time verifications invariably amount to a determination of space-time coincidences. If, for example, events consisted merely in the motion of material points, then ultimately nothing would be observable but the meetings of two or more of these points. Moreover, the results of our measurings are nothing but verifications of such meetings of the material points of our measuring instruments with other material points, coincidences between the hands of a clock and points on the clock dial, and observed point-events happening at the same place at the same time.

The introduction of a system of reference serves no other purpose than to facilitate the description of the totality of such coincidences. We allot to the universe four space-time variables x_1, x_2, x_3, x_4 in such a way that for every point-event there is a corresponding system of values of the variables $x_1 \ldots x_4$. To two coincident point-events there corresponds one system of values of the variables $x_1 \ldots x_4$, i.e. coincidence is characterized by the identity of the co-ordinates. If, in place of the variables $x_1 \ldots x_4$, we introduce functions of them, x'_1, x'_2, x'_3, x'_4, as a new system of co-ordinates, so that the systems of values are made to correspond to one another without ambiguity, the equality of all four co-ordinates in the new system will also serve as an expression for the space-time coincidence of the two point-events. As all our physical experience can be ultimately reduced to such coincidences, there is no immediate reason for preferring certain systems of co-ordinates to others, that is to say, we arrive at the requirement of general co-variance.

§ 4. THE RELATION OF THE FOUR CO-ORDINATES TO MEASUREMENT IN SPACE AND TIME

It is not my purpose in this discussion to represent the general theory of relativity as a system that is as simple and logical as possible, and with the minimum number of axioms; but my main object is to develop this theory in such a way that the reader will feel that the path we have entered upon is psychologically the natural one, and that the underlying assumptions will seem to have the highest possible degree of security. With this aim in view let it now be granted that:

For infinitely small four-dimensional regions the theory of relativity in the restricted sense is appropriate, if the co-ordinates are suitably chosen.

For this purpose we must choose the acceleration of the infinitely small ("local") system of co-ordinates so that no gravitational field occurs; this is possible for an infinitely small region. Let X_1, X_2, X_3, be the co-ordinates of space, and X_4 the appertaining co-ordinate of time measured in the appropriate unit.* If a rigid rod is imagined to be given as the unit measure, the co-ordinates, with a given orientation of the system of co-ordinates, have a direct physical meaning in the sense of the special theory of relativity. By the special theory of relativity the expression

$$ds^2 = -dX_1^2 - dX_2^2 - dX_3^2 + dX_4^2 \tag{1}$$

then has a value which is independent of the orientation of the local system of co-ordinates, and is ascertainable by measurements of space and time. The magnitude of the linear element pertaining to points of the four-dimensional continuum in infinite proximity, we call ds. If the ds belonging to the element $dX_1 \ldots dX_4$ is positive, we follow Minkowski in calling it time-like; if it is negative, we call it space-like.

To the "linear element" in question, or to the two infinitely proximate point-events, there will also correspond definite differentials $dx_1 \ldots dx_4$ of the four-dimensional co-ordinates of any chosen system of reference. If this system, as well as the "local" system, is given for the region under consideration, the dX_ν will allow themselves to be represented here by definite linear homogeneous expressions of the dx_σ:—

* The unit of time is to be chosen so that the velocity of light *in vacuo* as measured in the "local" system of co-ordinates is to be equal to unity.

$$dX_\nu = \sum_\sigma a_{\nu\sigma} dx_\sigma \qquad (2)$$

Inserting these expressions in (1), we obtain

$$ds^2 = \sum_{\tau\sigma} g_{\sigma\tau} dx_\sigma dx_\tau, \qquad (3)$$

where the $g_{\sigma\tau}$ will be functions of the x_σ. These can no longer be dependent on the orientation and the state of motion of the "local" system of co-ordinates, for ds^2 is a quantity ascertainable by rod-clock measurement of point-events infinitely proximate in space-time, and defined independently of any particular choice of co-ordinates. The $g_{\sigma\tau}$ are to be chosen here so that $g_{\sigma\tau} = g_{\tau\sigma}$; the summation is to extend over all values of σ and τ, so that the sum consists of 4 x 4 terms, of which twelve are equal in pairs.

The case of the ordinary theory of relativity arises out of the case here considered, if it is possible, by reason of the particular relations of the $g_{\sigma\tau}$ in a finite region, to choose the system of reference in the finite region in such a way that the $g_{\sigma\tau}$ assume the constant values

$$\left.\begin{matrix} -1 & 0 & 0 & 0 \\ 0 & -1 & 0 & 0 \\ 0 & 0 & -1 & 0 \\ 0 & 0 & 0 & +1 \end{matrix}\right\} \qquad (4)$$

We shall find hereafter that the choice of such co-ordinates is, in general, not possible for a finite region.

From the considerations of § 2 and § 3 it follows that the quantities $g_{\tau\sigma}$ are to be regarded from the physical standpoint as the quantities which describe the gravitational field in relation to the chosen system of reference. For, if we now assume the special theory of relativity to apply to a certain four-dimensional region with the co-ordinates properly chosen, then the $g_{\sigma\tau}$ have the values given in (4). A free material point then moves, relatively to this system, with uniform motion in a straight line. Then if we introduce new space-time co-ordinates x_1, x_2, x_3, x_4, by means of any substitution we choose, the $g_{\sigma\tau}$ in this new system will no longer be constants, but functions of space and time. At the same time the motion of the free material point will present itself in the new co-ordinates as a curvilinear non-uniform motion, and the law of this motion will be independent of the nature of the moving particle. We shall therefore interpret this motion as a motion under the influence of a gravitational field. We thus find the occurrence of a gravitational field connected with a space-time variability of the g_σ. So, too, in the general case, when we are no longer able by a suitable choice of co-ordinates to apply the special theory of relativity to a finite region, we shall hold fast to the view that the $g_{\sigma\tau}$ describe the gravitational field.

Thus, according to the general theory of relativity, gravitation occupies an exceptional position with regard to other forces, particularly the electromagnetic forces, since the ten functions representing the gravitational field at the same time define the metrical properties of the space measured.

B. MATHEMATICAL AIDS TO THE FORMULATION OF GENERALLY COVARIANT EQUATIONS

Having seen in the foregoing that the general postulate of relativity leads to the requirement that the equations of physics shall be covariant in the face of any substitution of the co-ordinates $x_1 \ldots x_4$, we have to consider how such generally covariant equations can be found. We now turn to this purely mathematical task, and we shall find that in its solution a fundamental rôle is played by the invariant ds given in equation (3), which, borrowing from Gauss's theory of surfaces, we have called the "linear element."

The fundamental idea of this general theory of covariants is the following:—Let certain things ("tensors") be defined with respect to any system of co-ordinates by a number of functions of the co-ordinates, called the "components" of the tensor. There are then certain rules by which these components can be calculated for a new system of co-ordinates, if they are known for the original system of co-ordinates, and if the transformation connecting the two systems is known. The things hereafter called tensors are further characterized by the fact that the equations of transformation for their components are linear and homogeneous. Accordingly, all the components in the new system vanish, if they all vanish in the original system. If, therefore, a law of nature is expressed by equating all the components of a tensor to zero, it is generally covariant. By examining the laws of the formation of tensors, we acquire the means of formulating generally covariant laws.

§ 5. CONTRAVARIANT AND COVARIANT FOUR-VECTORS

Contravariant Four-vectors.—The linear element is defined by the four "components" dx_ν, for which the law of transformation is expressed by the equation

$$dx'_\sigma = \sum_\nu \frac{\partial x'_\sigma}{\partial x_\nu} dx_\nu \qquad (5)$$

The dx'_σ are expressed as linear and homogeneous functions of the dx_ν. Hence we may look upon these co-ordinate differentials as the components of a "tensor" of the particular kind which we call a contravariant four-vector. Any thing which is defined relatively to the system of co-ordinates by four quantities A^ν, and which is transformed by the same law

$$A'^\sigma = \sum_\nu \frac{\partial x'_\sigma}{\partial x_\nu} A^\nu, \tag{5a}$$

we also call a contravariant four-vector. From (5a) it follows at once that the sums $A^\sigma \pm B^\sigma$ are also components of a four-vector, if A^σ and B^σ are such. Corresponding relations hold for all "tensors" subsequently to be introduced. (Rule for the addition and subtraction of tensors.)

Covariant Four-vectors.—We call four quantities $A\nu$ the components of a covariant four-vector, if for any arbitrary choice of the contravariant four-vector B^ν

$$\sum_\nu A_\nu B^\nu = \text{Invariant} \tag{6}$$

The law of transformation of a covariant four-vector follows from this definition. For if we replace B^ν on the right-hand side of the equation

$$\sum_\sigma A'_\sigma B'^\sigma = \sum_\nu A_\nu B^\nu$$

by the expression resulting from the inversion of (5a),

$$\sum_\sigma \frac{\partial x_\nu}{\partial x'_\sigma} B'^\sigma,$$

we obtain

$$\sum_\sigma B'^\sigma \sum_\nu \frac{\partial x_\nu}{\partial x'_\sigma} A_\nu = \sum_\sigma B'^\sigma A'_\sigma.$$

Since this equation is true for arbitrary values of the B'^σ, it follows that the law of transformation is

$$A'_\sigma = \sum_\nu \frac{\partial x_\nu}{\partial x'_\sigma} A_\nu. \tag{7}$$

Note on a Simplified Way of Writing the Expressions.—A glance at the equations of this paragraph shows that there is always a summation with respect to the indices which occur twice under a sign of summation (e.g. the index ν in (5)), and only with respect to indices which occur twice. It is therefore possible, without loss of clearness, to omit the sign of summation. In its place we introduce the convention:—If an index occurs twice in one term of an expression, it is always to be summed unless the contrary is expressly stated.

The difference between covariant and contravariant four-vectors lies in the law of transformation ((7) or (5) respectively). Both forms are tensors in the sense of the general remark above. Therein lies their importance. Following Ricci and Levi-Civita, we denote the contravariant character by placing the index above, the covariant by placing it below.

§ 6. TENSORS OF THE SECOND AND HIGHER RANKS

Contravariant Tensors.—If we form all the sixteen products $A^{\mu\nu}$ of the components A^{μ} and B^{ν} of two contravariant four-vectors

$$A^{\mu\nu} = A^{\mu}B^{\nu} \tag{8}$$

then by (8) and (5a) $A^{\mu\nu}$ satisfies the law of transformation

$$A'^{\sigma\tau} = \frac{\partial x'_{\sigma}}{\partial x_{\mu}} \frac{\partial x'_{\tau}}{\partial x_{\nu}} A^{\mu\nu} \tag{9}$$

We call a thing which is described relatively to any system of reference by sixteen quantities, satisfying the law of transformation (9), a contravariant tensor of the second rank. Not every such tensor allows itself to be formed in accordance with (8) from two four-vectors, but it is easily shown that any given sixteen $A^{\mu\nu}$ can be represented as the sums of the $A^{\mu}B^{\nu}$ of four appropriately selected pairs of four-vectors. Hence we can prove nearly all the laws which apply to the tensor of the second rank defined by (9) in the simplest manner by demonstrating them for the special tensors of the type (8).

Contravariant Tensors of Any Rank.—It is clear that, on the lines of (8) and (9), contravariant tensors of the third and higher ranks may also be defined with 4^3 components, and so on. In the same way it follows from (8) and (9) that the contravariant four-vector may be taken in this sense as a contravariant tensor of the first rank.

Covariant Tensors.—On the other hand, if we take the sixteen products $A_{\mu\nu}$, of two covariant four-vectors A_{μ} and B_{ν},

$$A_{\mu\nu} = A_{\mu}B_{\nu}, \tag{10}$$

the law of transformation for these is

$$A'_{\sigma\tau} = \frac{\partial x_{\mu}}{\partial x'_{\sigma}} \frac{\partial x_{\nu}}{\partial x'_{\tau}} A_{\mu\nu} \tag{11}$$

This law of transformation defines the covariant tensor of the second rank. All our previous remarks on contravariant tensors apply equally to covariant tensors.

NOTE.—It is convenient to treat the scalar (or invariant) both as a contravariant and a covariant tensor of zero rank.

Mixed Tensors.—We may also define a tensor of the second rank of the type

$$A_{\mu}^{\nu} = A_{\mu}B^{\nu}$$

which is covariant with respect to the index μ, and contravariant with respect to the index ν. Its law of transformation is

$$A'^{\tau}_{\sigma} = \frac{\partial x'_{\tau}}{\partial x_{\nu}} \frac{\partial x_{\mu}}{\partial x'_{\sigma}} A^{\nu}_{\mu} \tag{13}$$

Naturally there are mixed tensors with any number of indices of covariant character, and any number of indices of contravariant character. Covariant and contravariant tensors may be looked upon as special cases of mixed tensors.

Symmetrical Tensors.—A contravariant, or a covariant tensor, of the second or higher rank is said to be symmetrical if two components, which are obtained the one from the other by the interchange of two indices, are equal. The tensor $A^{\mu\nu}$, or the tensor $A_{\mu\nu}$, is thus symmetrical if for any combination of the indices μ, ν,

$$A^{\mu\nu} = A^{\nu\mu}, \tag{14}$$

or respectively,

$$A_{\mu\nu} = A_{\nu\mu}. \tag{14a}$$

It has to be proved that the symmetry thus defined is a property which is independent of the system of reference. It follows in fact from (9), when (14) is taken into consideration, that

$$A'^{\sigma\tau} = \frac{\partial x'_{\sigma}}{\partial x_{\mu}} \frac{\partial x'_{\tau}}{\partial x_{\nu}} A^{\mu\nu} = \frac{\partial x'_{\sigma}}{\partial x_{\mu}} \frac{\partial x'_{\tau}}{\partial x_{\nu}} A^{\nu\mu} = \frac{\partial x'_{\sigma}}{\partial x_{\nu}} \frac{\partial x'_{\tau}}{\partial x_{\mu}} A^{\mu\nu} = A'^{\tau\sigma}$$

The last equation but one depends upon the interchange of the summation indices μ and ν, i.e. merely on a change of notation.

Antisymmetrical Tensors.—A contravariant or a covariant tensor of the second, third, or fourth rank is said to be antisymmetrical if two components, which are obtained the one from the other by the interchange of two indices, are equal and of opposite sign. The tensor $A^{\mu\nu}$, or the tensor $A_{\mu\nu}$, is therefore antisymmetrical, if always

$$A^{\mu\nu} = -A^{\nu\mu}, \tag{15}$$

or respectively,

$$A_{\mu\nu} = -A_{\nu\mu}. \tag{15a}$$

Of the sixteen components $A^{\mu\nu}$, the four components $A^{\mu\mu}$ vanish; the rest are equal and of opposite sign in pairs, so that there are only six components numerically different (a six-vector). Similarly we see that the antisymmetrical tensor of the third rank $A^{\mu\nu\sigma}$ has only four numerically different components, while the antisymmetrical tensor $A^{\mu\nu\sigma\tau}$ has only one. There are no antisymmetrical tensors of higher rank than the fourth in a continuum of four dimensions.

§ 7. MULTIPLICATION OF TENSORS

Outer Multiplication of Tensors.—We obtain from the components of a tensor of rank n and of a tensor of rank m the components of a tensor of rank $n + m$ by multiplying each component of the one tensor by each component of the other. Thus, for example, the tensors T arise out of the tensors A and B of different kinds,

$$T_{\mu\nu\sigma} = A_{\mu\nu}B_{\sigma},$$
$$T^{\mu\nu\sigma\tau} = A^{\mu\nu}B^{\sigma\tau},$$
$$T^{\sigma\tau}_{\mu\nu} = A_{\mu\nu}B^{\sigma\tau}.$$

The proof of the tensor character of T is given directly by the representations (8), (10), (12), or by the laws of transformation (9), (11), (13). The equations (8), (10), (12) are themselves examples of outer multiplication of tensors of the first rank.

"Contraction" of a Mixed Tensor.—From any mixed tensor we may form a tensor whose rank is less by two, by equating an index of covariant with one of contravariant character, and summing with respect to this index ("contraction"). Thus, for example, from the mixed tensor of the fourth rank $A^{\sigma\tau}_{\mu\nu}$, we obtain the mixed tensor of the second rank,

$$A^{\tau}_{\nu} = A^{\mu\tau}_{\mu\nu} \left(= \sum_{\mu} A^{\mu\tau}_{\mu\nu} \right),$$

and from this, by a second contraction, the tensor of zero rank,

$$A = A^{\nu}_{\nu} = A^{\mu\nu}_{\mu\nu}.$$

The proof that the result of contraction really possesses the tensor character is given either by the representation of a tensor according to the generalization of (12) in combination with (6), or by the generalization of (13).

Inner and Mixed Multiplication of Tensors.—These consist in a combination of outer multiplication with contraction.

Examples.—From the covariant tensor of the second rank $A_{\mu\nu}$ and the contravariant tensor of the first rank B^{σ} we form by outer multiplication the mixed tensor

$$D^{\sigma}_{\mu\nu} = A_{\mu\nu}B^{\sigma}.$$

On contraction with respect to the indices ν and σ, we obtain the covariant four-vector

$$D_{\mu} = D^{\nu}_{\mu\nu} = A_{\mu\nu}B^{\nu}.$$

This we call the inner product of the tensors $A_{\mu\nu}$ and B^{σ}. Analogously we form from the tensors $A_{\mu\nu}$, and $B^{\sigma\tau}$, by outer multiplication and double contraction, the inner product $A_{\mu\nu}B^{\mu\nu}$. By outer multiplication and one contraction, we obtain from $A\mu\nu$ and $B^{\sigma\tau}$ the mixed tensor of the second rank $D_{\mu}^{\tau} = A_{\mu\nu}B^{\nu\tau}$. This operation may be aptly characterized as a mixed one, being "outer" with respect to the indices μ and τ, and "inner" with respect to the indices ν and σ.

We now prove a proposition which is often useful as evidence of tensor character. From what has just been explained, $A_{\mu\nu}B^{\mu\nu}$ is a scalar if $A_{\mu\nu}$, and $B^{\sigma\tau}$ are tensors. But we may also make the following assertion: If $A_{\mu\nu}B^{\mu\nu}$ is a scalar *for any choice of the tensor* $B^{\mu\nu}$, then $A_{\mu\nu}$ has tensor character. For, by hypothesis, for any substitution,

$$A'_{\sigma\tau} B'^{\sigma\tau} = A_{\mu\nu} B^{\mu\nu}.$$

But by an inversion of (9)

$$B^{\mu\nu} = \frac{\partial x_\mu}{\partial x'_\sigma} \frac{\partial x_\nu}{\partial x'_\tau} B'^{\sigma\tau}.$$

This, inserted in the above equation, gives

$$\left(A'_{\sigma\tau} - \frac{\partial x_\mu}{\partial x'_\sigma} \frac{\partial x_\nu}{\partial x'_\tau} A_{\mu\nu} \right) B'^{\sigma\tau} = 0.$$

This can only be satisfied for arbitrary values of $B'^{\sigma\tau}$ if the bracket vanishes. The result then follows by equation (11). This rule applies correspondingly to tensors of any rank and character, and the proof is analogous in all cases.

The rule may also be demonstrated in this form: If B^{μ} and C^{ν} are any vectors, and if, for all values of these, the inner product $A_{\mu\nu}B^{\mu}C^{\nu}$ is a scalar, then $A_{\mu\nu}$ is a covariant tensor. This latter proposition also holds good even if only the more special assertion is correct, that with any choice of the four-vector B^{μ} the inner product $A_{\mu\nu}B^{\mu}B^{\nu}$ is a scalar, if in addition it is known that $A_{\mu\nu}$ satisfies the condition of symmetry $A_{\mu\nu} = A_{\nu\mu}$. For by the method given above we prove the tensor character of $(A_{\mu\nu} + A_{\nu\mu})$, and from this the tensor character of $A_{\mu\nu}$ follows on account of symmetry. This also can be easily generalized to the case of covariant and contravariant tensors of any rank.

Finally, there follows from what has been proved, this law, which may also be generalized for any tensors: If for any choice of the four-vector B^{ν} the quantities $A_{\mu\nu}B^{\nu}$ form a tensor of the first rank, then $A_{\mu\nu}$ is a tensor of the second rank. For, if C^{μ} is any four-vector, then on account of the tensor character of $A_{\mu\nu}B^{\nu}$, the inner product $A_{\mu\nu}B^{\nu}C^{\mu}$ is a scalar for any choice of the two four-vectors B^{ν} and C^{μ}. From which the proposition follows.

§ 8. SOME ASPECTS OF THE FUNDAMENTAL TENSOR $g_{\mu\nu}$

The Covariant Fundamental Tensor.—In the invariant expression for the square of the linear element,

$$ds^2 = g_{\mu\nu} dx_\mu dx_\nu,$$

the part played by the dx_μ is that of a contravariant vector which may be chosen at will. Since further, $g_{\mu\nu} = g_{\nu\mu}$, it follows from the considerations of the preceding paragraph that $g_{\mu\nu}$ is a covariant tensor of the second rank. We call it the "fundamental tensor." In what follows we deduce some properties of this tensor which, it is true, apply to any tensor of the second rank. But as the fundamental tensor plays a special part in our theory, which has its physical basis in the peculiar effects of gravitation, it so happens that the relations to be developed are of importance to us only in the case of the fundamental tensor.

The Contravariant Fundamental Tensor.—If in the determinant formed by the elements $g_{\mu\nu}$, we take the co-factor of each of the $g_{\mu\nu}$ and divide it by the determinant $g = |g_{\mu\nu}|$, we obtain certain quantities $g^{\mu\nu} (= g^{\nu\mu})$ which, as we shall demonstrate, form a contravariant tensor.

By a known property of determinants

$$g_{\mu\sigma} g^{\nu\sigma} = \delta_\mu^\nu \tag{16}$$

where the symbol δ_μ^ν denotes 1 or 0, according as $\mu = \nu$ or $\mu \neq \nu$.

Instead of the above expression for ds^2 we may thus write

$$g_{\mu\sigma} \delta_\nu^\sigma dx_\mu dx_\nu$$

or, by (16)

$$g_{\mu\sigma} g_{\nu\tau} g^{\sigma\tau} dx_\mu dx_\nu$$

But, by the multiplication rules of the preceding paragraphs, the quantities

$$d\xi_\sigma = g_{\mu\sigma} dx_\mu$$

form a covariant four-vector, and in fact an arbitrary vector, since the dx_μ are arbitrary. By introducing this into our expression we obtain

$$ds^2 = g^{\sigma\tau} d\xi_\sigma d\xi_\tau$$

Since this, with the arbitrary choice of the vector $d\xi_\sigma$, is a scalar, and $g^{\sigma\tau}$ by its definition is symmetrical in the indices σ and τ, it follows from the results of the preceding paragraph that $g^{\sigma\tau}$ is a contravariant tensor.

It further follows from (16) that δ_μ^ν is also a tensor, which we may call the mixed fundamental tensor.

The Determinant of the Fundamental Tensor.—By the rule for the multiplication of determinants

$$\left|g_{\mu\alpha}g^{\alpha\nu}\right| = \left|g_{\mu\alpha}\right| \times \left|g^{\alpha\nu}\right|$$

On the other hand

$$\left|g_{\mu\alpha}g^{\alpha\nu}\right| = \left|\delta_{\mu}^{\nu}\right| = 1.$$

It therefore follows that

$$\left|g_{\mu\nu}\right| \times \left|g^{\mu\nu}\right| = 1 \qquad (17)$$

The Volume Scalar.—We seek first the law of transformation of the determinant $g = \left|g_{\mu\nu}\right|$. In accordance with (11)

$$g' = \left|\frac{\partial x_{\mu}}{\partial x'_{\sigma}}\frac{\partial x_{\nu}}{\partial x'_{\tau}}g_{\mu\nu}\right|.$$

Hence, by a double application of the rule for the multiplication of determinants, it follows that

$$g' = \left|\frac{\partial x_{\mu}}{\partial x'_{\sigma}}\right| \cdot \left|\frac{\partial x_{\nu}}{\partial x'_{\tau}}\right| \cdot \left|g_{\mu\nu}\right| = \left|\frac{\partial x_{\mu}}{\partial x'_{\sigma}}\right|^{2} g,$$

or

$$\sqrt{g'} = \left|\frac{\partial x_{\mu}}{\partial x'_{\sigma}}\right| = \sqrt{g}.$$

On the other hand, the law of transformation of the element of volume

$$d\tau = \int dx_1 dx_2 dx_3 dx_4$$

is, in accordance with the theorem of Jacobi,

$$d\tau' = \left|\frac{\partial x'_{\sigma}}{\partial x_{\mu}}\right| d\tau.$$

By multiplication of the last two equations, we obtain

$$\sqrt{g'}d\tau' = \sqrt{g}d\tau \qquad (18).$$

Instead of $\sqrt{g}$, we introduce in what follows the quantity $\sqrt{-g}$, which is always real on account of the hyperbolic character of the space-time continuum. The invariant $\sqrt{-g}\,d\tau$ is equal to the magnitude of the four-dimensional element of volume in the "local" system of reference, as measured with rigid rods and clocks in the sense of the special theory of relativity.

Note on the Character of the Space-time Continuum.—Our assumption that the special theory of relativity can always be applied to an infinitely small region, implies that ds^2 can always be expressed in accordance with (1) by means of real quantities $dX_1 \ldots dX_4$. If we denote by $d\tau_0$ the "natural" element of volume dX_1, dX_2, dX_3, dX_4, then

$$d\tau_0 = \sqrt{-g}\, d\tau \tag{18a}$$

If $\sqrt{-g}$ were to vanish at a point of the four-dimensional continuum, it would mean that at this point an infinitely small "natural" volume would correspond to a finite volume in the co-ordinates. Let us assume that this is never the case. Then g cannot change sign. We will assume that, in the sense of the special theory of relativity, g always has a finite negative value. This is a hypothesis as to the physical nature of the continuum under consideration, and at the same time a convention as to the choice of co-ordinates.

But if $-g$ is always finite and positive, it is natural to settle the choice of co-ordinates *a posteriori* in such a way that this quantity is always equal to unity. We shall see later that by such a restriction of the choice of co-ordinates it is possible to achieve an important simplification of the laws of nature.

In place of (18), we then have simply $d\tau' = d\tau$, from which, in view of Jacobi's theorem, it follows that

$$\left| \frac{\partial x'_\sigma}{\partial x_\mu} \right| = 1 \tag{19}$$

Thus, with this choice of co-ordinates, only substitutions for which the determinant is unity are permissible.

But it would be erroneous to believe that this step indicates a partial abandonment of the general postulate of relativity. We do not ask "What are the laws of nature which are covariant in face of all substitutions for which the determinant is unity?" but our question is "What are the generally covariant laws of nature?" It is not until we have formulated these that we simplify their expression by a particular choice of the system of reference.

The Formation of New Tensors by Means of the Fundamental Tensor.—Inner, outer, and mixed multiplication of a tensor by the fundamental tensor give tensors of different character and rank. For example,

$$A^\mu = g^{\mu\sigma} A_\sigma,$$
$$A = g_{\mu\nu} A^{\mu\nu}.$$

The following forms may be specially noted:—

$$A^{\mu\nu} = g^{\mu\alpha} g^{\nu\beta} A_{\alpha\beta},$$
$$A_{\mu\nu} = g_{\mu\alpha} g_{\nu\beta} A^{\alpha\beta}$$

(the "complements" of covariant and contravariant tensors respectively), and

$$B_{\mu\nu} = g_{\mu\nu} g^{\alpha\beta} A_{\alpha\beta}$$

We call $B_{\mu\nu}$ the reduced tensor associated with $A_{\mu\nu}$. Similarly,

$$B^{\mu\nu} = g^{\mu\nu} g_{\alpha\beta} A^{\alpha\beta}.$$

It may be noted that $g^{\mu\nu}$ is nothing more then the complement of $g_{\mu\nu}$, since

$$g^{\mu\alpha} g^{\nu\beta} g_{\alpha\beta} = g^{\mu\alpha} \delta^{\nu}_{\alpha} = g^{\mu\nu}.$$

§ 9. THE EQUATION OF THE GEODETIC LINE. THE MOTION OF A PARTICLE

As the linear element ds is defined independently of the system of co-ordinates, the line drawn between two points P and P' of the four-dimensional continuum in such a way that $\int ds$ is stationary—a geodetic line—has a meaning which also is independent of the choice of co-ordinates. Its equation is

$$\delta \int_{P}^{P'} ds = 0 \tag{20}$$

Carrying out the variation in the usual way, we obtain from this equation four differential equations which define the geodetic line; this operation will be inserted here for the sake of completeness. Let λ be a function of the co-ordinates x_ν, and let this define a family of surfaces which intersect the required geodetic line as well as all the lines in immediate proximity to it which are drawn through the points P and P'. Any such line may then be supposed to be given by expressing its co-ordinates x_ν as functions of λ. Let the symbol δ indicate the transition from a point of the required geodetic to the point corresponding to the same λ on a neighbouring line. Then for (20) we may substitute

$$\left.\begin{aligned} &\int_{\lambda_1}^{\lambda_2} \delta w \, d\lambda = 0 \\ &w^2 = g_{\mu\nu} \frac{dx_\mu}{d\lambda} \frac{dx_\nu}{d\lambda} \end{aligned}\right\} \tag{20a}$$

But since

$$\delta w = \frac{1}{w} \left\{ \frac{1}{2} \frac{\partial g_{\mu\nu}}{\partial x_\sigma} \frac{dx_\mu}{d\lambda} \frac{dx_\nu}{d\lambda} \delta x_\sigma + g_{\mu\nu} \frac{dx_\mu}{d\lambda} \delta \left(\frac{dx_\nu}{d\lambda} \right) \right\},$$

and

$$\delta \left(\frac{dx_\nu}{d\lambda} \right) = \frac{d}{d\lambda} (\delta x_\nu),$$

we obtain from (20a), after a partial integration,

$$\int_{\lambda_1}^{\lambda_2} \kappa_\sigma \delta x_\sigma d\lambda = 0,$$

where

$$\kappa_\sigma = \frac{d}{d\lambda}\left\{\frac{g_{\mu\nu}}{w}\frac{dx_\mu}{d\lambda}\right\} - \frac{1}{2w}\frac{\partial g_{\mu\nu}}{\partial x_\sigma}\frac{dx_\mu}{d\lambda}\frac{dx_\nu}{d\lambda} \tag{20b}$$

Since the values of δx_σ are arbitrary, it follows from this that

$$\kappa_\sigma = 0 \tag{20c}$$

are the equations of the geodetic line.

If *ds* does not vanish along the geodetic line we may choose the "length of the arc" *s*, measured along the geodetic line, for the parameter λ. Then $w = 1$, and in place of (20c) we obtain

$$g_{\mu\nu}\frac{d^2 x_\mu}{ds^2} + \frac{\partial g_{\mu\nu}}{\partial x_\sigma}\frac{dx_\sigma}{ds}\frac{dx_\mu}{ds} - \frac{1}{2}\frac{\partial g_{\mu\nu}}{\partial x_\sigma}\frac{dx_\mu}{ds}\frac{dx_\nu}{ds} = 0$$

or, by a mere change of notation,

$$g_{\alpha\sigma}\frac{d^2 x_\alpha}{ds^2} + [\mu\nu, \sigma]\frac{dx_\mu}{ds}\frac{dx_\nu}{ds} = 0 \tag{20d}$$

where, following Christoffel, we have written

$$[\mu\nu, \sigma] = \frac{1}{2}\left(\frac{\partial g_{\mu\sigma}}{\partial x_\nu} + \frac{\partial g_{\nu\sigma}}{\partial x_\mu} - \frac{\partial g_{\mu\nu}}{\partial x_\sigma}\right) \tag{21}$$

Finally, if we multiply (20d) by $g^{\sigma\tau}$ (outer multiplication with respect to τ, inner with respect to σ), we obtain the equations of the geodetic line in the form

$$\frac{d^2 x_\tau}{ds^2} + \{\mu\nu, \tau\}\frac{dx_\mu}{ds}\frac{dx_\nu}{ds} = 0 \tag{22}$$

where, following Christoffel, we have set

$$\{\mu\nu, \tau\} = g^{\tau\alpha}[\mu\nu, \alpha] \tag{23}$$

§ 10. THE FORMATION OF TENSORS BY DIFFERENTIATION

With the help of the equation of the geodetic line we can now easily deduce the laws by which new tensors can be formed from old by differentiation. By this means we are able for the first time to formulate generally covariant differential equations. We reach this goal by repeated application of the following simple law:

If in our continuum a curve is given, the points of which are specified by the arcual distance *s* measured from a fixed point on the curve, and if, further, ϕ is an invariant function of space, then $d\phi/ds$ is also an invariant. The proof lies in this, that *ds* is an invariant as well as $d\phi$.

As

$$\frac{d\phi}{ds} = \frac{\partial \phi}{\partial x_\mu}\frac{dx_\mu}{ds}$$

therefore

$$\psi = \frac{\partial \phi}{\partial x_\mu} \frac{dx_\mu}{ds}$$

is also an invariant, and an invariant for all curves starting from a point of the continuum, that is, for any choice of the vector dx_μ. Hence it immediately follows that

$$A_\mu = \frac{\partial \phi}{\partial x_\mu} \qquad (24)$$

is a covariant four-vector—the "gradient" of ϕ.

According to our rule, the differential quotient

$$\chi = \frac{d\psi}{ds}$$

taken on a curve, is similarly an invariant. Inserting the value of ψ, we obtain in the first place

$$\chi = \frac{\partial^2 \phi}{\partial x_\mu \partial x_\nu} \frac{dx_\mu}{ds} \frac{dx_\nu}{ds} + \frac{\partial \phi}{\partial x_\mu} \frac{d^2 x_\mu}{ds^2}$$

The existence of a tensor cannot be deduced from this forthwith. But if we may take the curve along which we have differentiated to be a geodetic, we obtain on substitution for d^2x_ν / ds^2 from (22),

$$\chi = \left(\frac{\partial^2 \phi}{\partial x_\mu \partial x_\nu} - \{\mu\nu, \tau\} \frac{\partial \phi}{\partial x_\tau} \right) \frac{dx_\mu}{ds} \frac{dx_\nu}{ds}.$$

Since we may interchange the order of the differentiations, and since by (23) and (21) $\{\mu\nu, \tau\}$ is symmetrical in μ and ν, it follows that the expression in brackets is symmetrical in μ and ν. Since a geodetic line can be drawn in any direction from a point of the continuum, and therefore dx_μ / ds is a four-vector with the ratio of its components arbitrary, it follows from the results of § 7 that

$$A_{\mu\nu} = \frac{\partial^2 \phi}{\partial x_\mu \partial x_\nu} - \{\mu\nu, \tau\} \frac{\partial \phi}{\partial x_\tau} \qquad (25)$$

is a covariant tensor of the second rank. We have therefore come to this result: from the covariant tensor of the first rank

$$A_\mu = \frac{\partial \phi}{\partial x_\mu}$$

we can, by differentiation, form a covariant tensor of the second rank

$$A_{\mu\nu} = \frac{\partial A_\mu}{\partial x_\nu} - \{\mu\nu, \tau\} A_\tau \qquad (26)$$

We call the tensor $A_{\mu\nu}$ the "extension" (covariant derivative) of the tensor A_μ. In the first place we can readily show that the operation leads to a tensor, even if the vector A_μ cannot be represented as a gradient. To see this, we first observe that

$$\psi \frac{\partial \phi}{\partial x_\mu}$$

is a covariant vector, if ψ and ϕ are scalars. The sum of four such terms

$$S_\mu = \psi^{(1)} \frac{\partial \phi^{(1)}}{\partial x_\mu} + . + . + \psi^{(4)} \frac{\partial \phi^{(4)}}{\partial x_\mu},$$

is also a covariant vector, if $\psi^{(1)}$, $\phi^{(1)} \ldots \psi^{(4)}$, $\phi^{(4)}$ are scalars. But it is clear that any covariant vector can be represented in the form S_μ. For, if A_μ is a vector whose components are any given functions of the x_ν, we have only to put (in terms of the selected system of co-ordinates)

$$\begin{aligned} \psi^{(1)} &= A_1, & \phi^{(1)} &= x_1, \\ \psi^{(2)} &= A_2, & \phi^{(2)} &= x_2, \\ \psi^{(3)} &= A_3, & \phi^{(3)} &= x_3, \\ \psi^{(4)} &= A_4, & \phi^{(4)} &= x_4, \end{aligned}$$

in order to ensure that S_μ shall be equal to A_μ.

Therefore, in order to demonstrate that $A_{\mu\nu}$ is a tensor if *any* covariant vector is inserted on the right-hand side for A_μ, we only need show that this is so for the vector S_μ. But for this latter purpose it is sufficient, as a glance at the right-hand side of (26) teaches us, to furnish the proof for the case

$$A_\mu = \psi \frac{\partial \phi}{\partial x_\mu}.$$

Now the right-hand side of (25) multiplied by ψ,

$$\psi \frac{\partial^2 \phi}{\partial x_\mu \partial x_\nu} - \{\mu\nu, \tau\} \psi \frac{\partial \phi}{\partial x_\tau}$$

is a tensor. Similarly

$$\frac{\partial \psi}{\partial x_\mu} \frac{\partial \phi}{\partial x_\nu}$$

being the outer product of two vectors, is a tensor. By addition, there follows the tensor character of

$$\frac{\partial}{\partial x_\nu}\left(\psi \frac{\partial \phi}{\partial x_\mu}\right) - \{\mu\nu, \tau\}\left(\psi \frac{\partial \phi}{\partial x_\tau}\right).$$

As a glance at (26) will show, this completes the demonstration for the vector

$$\psi \frac{\partial \phi}{\partial x_\mu}$$

and consequently, from what has already been proved, for any vector A_μ.

By means of the extension of the vector, we may easily define the "extension" of a covariant tensor of any rank. This operation is a generalization of the extension of a vector. We restrict ourselves to the case of a tensor of the second rank, since this suffices to give a clear idea of the law of formation.

As has already been observed, any covariant tensor of the second rank can be represented* as the sum of tensors of the type $A_\mu B_\nu$. It will therefore be sufficient to deduce the expression for the extension of a tensor of this special type. By (26) the expressions

$$\frac{\partial A_\mu}{\partial x_\sigma} - \{\sigma\mu, \tau\} A_\tau,$$

$$\frac{\partial B_\nu}{\partial x_\sigma} - \{\sigma\nu, \tau\} B_\tau,$$

are tensors. On outer multiplication of the first by B_ν, and of the second by A_μ, we obtain in each case a tensor of the third rank. By adding these, we have the tensor of the third rank

$$A_{\mu\nu\sigma} = \frac{\partial A_{\mu\nu}}{\partial x_\sigma} - \{\sigma\mu, \tau\} A_{\tau\nu} - \{\sigma\nu, \tau\} A_{\mu\tau} \qquad (27)$$

where we have put $A_{\mu\nu} = A_\mu B_\nu$. As the right-hand side of (27) is linear and homogeneous in the $A_{\mu\nu}$ and their first derivatives, this law of formation leads to a tensor, not only in the case of a tensor of the type $A_\mu B_\nu$, but also in the case of a sum of such tensors, i.e. in the case of any covariant tensor of the second rank. We call $A_{\mu\nu\sigma}$ the extension of the tensor $A_{\mu\nu}$.

It is clear that (26) and (24) concern only special cases of extension (the extension of the tensors of rank one and zero respectively).

In general, all special laws of formation of tensors are included in (27) in combination with the multiplication of tensors.

§ 11. Some Cases of Special Importance

The Fundamental Tensor.—We will first prove some lemmas which will be useful hereafter. By the rule for the differentiation of determinants

$$dg = g^{\mu\nu} g dg_{\mu\nu} = -g_{\mu\nu} g dg^{\mu\nu} \qquad (28)$$

The last member is obtained from the last but one, if we bear in mind that $g_{\mu\nu} g^{\mu'\nu} = \delta_\mu^{\mu'}$, so that $g_{\mu\nu} g^{\mu\nu} = 4$, and consequently

$$g_{\mu\nu} dg^{\mu\nu} + g^{\mu\nu} dg_{\mu\nu} = 0.$$

From (28), it follows that

* By outer multiplication of the vector with arbitrary components $A_{11}, A_{12}, A_{13}, A_{14}$ by the vector with components 1, 0, 0, 0, we produce a tensor with components

A_{11}	A_{12}	A_{13}	A_{14}
0	0	0	0
0	0	0	0
0	0	0	0

By the addition of four tensors of this type, we obtain the tensor $A_{\mu\nu}$ with any assigned components.

$$\frac{1}{\sqrt{-g}}\frac{\partial\sqrt{-g}}{\partial x_\sigma}=\frac{1}{2}\frac{\partial\log(-g)}{\partial x_\sigma}=\frac{1}{2}g^{\mu\nu}\frac{\partial g_{\mu\nu}}{\partial x_\sigma}=\frac{1}{2}g_{\mu\nu}\frac{\partial g^{\mu\nu}}{\partial x_\sigma} \tag{29}$$

Further, from $g_{\mu\sigma}g^{\nu\sigma}=\delta_\mu^\nu$, it follows on differentiation that

$$\left.\begin{aligned} g_{\mu\sigma}dg^{\nu\sigma} &= -g^{\nu\sigma}dg_{\mu\sigma} \\ g_{\mu\sigma}\frac{\partial g^{\nu\sigma}}{\partial x_\lambda} &= -g^{\nu\sigma}d\frac{\partial g_{\mu\sigma}}{\partial x_\lambda} \end{aligned}\right\} \tag{30}$$

From these, by mixed multiplication by $g^{\sigma\tau}$ and $g_{\nu\lambda}$ respectively, and a change of notation for the indices, we have

$$\left.\begin{aligned} dg^{\mu\nu} &= -g^{\mu\alpha}g^{\nu\beta}dg_{\alpha\beta} \\ \frac{\partial g^{\mu\nu}}{\partial x_\sigma} &= -g^{\mu\alpha}g^{\nu\beta}\frac{\partial g_{\alpha\beta}}{\partial x_\sigma} \end{aligned}\right\} \tag{31}$$

and

$$\left.\begin{aligned} dg_{\mu\nu} &= -g_{\mu\alpha}g_{\nu\beta}dg^{\alpha\beta} \\ \frac{\partial g_{\mu\nu}}{\partial x_\sigma} &= -g_{\mu\alpha}g_{\nu\beta}\frac{\partial g^{\alpha\beta}}{\partial x_\sigma} \end{aligned}\right\} \tag{32}$$

The relation (31) admits of a transformation, of which we also have frequently to make use. From (21)

$$\frac{\partial g_{\alpha\beta}}{\partial x_\sigma}=[\alpha\sigma,\beta]+[\beta\sigma,\alpha] \tag{33}$$

Inserting this in the second formula of (31), we obtain, in view of (23)

$$\frac{\partial g^{\mu\nu}}{\partial x_\sigma}=-g^{\mu\tau}\{\tau\sigma,\nu\}-g^{\nu\tau}\{\tau\sigma,\mu\} \tag{34}$$

Substituting the right-hand side of (34) in (29), we have

$$\frac{1}{\sqrt{-g}}\frac{\partial\sqrt{-g}}{\partial x_\sigma}=\{\mu\sigma,\mu\} \tag{29a}$$

The "Divergence" of a Contravariant Vector.—If we take the inner product of (26) by the contravariant fundamental tensor $g^{\mu\nu}$, the right-hand side, after a transformation of the first term, assumes the form

$$\frac{\partial}{\partial x_\nu}\left(g^{\mu\nu}A_\mu\right)-A_\mu\frac{\partial g^{\mu\nu}}{\partial x_\nu}-\frac{1}{2}g^{\tau\alpha}\left(\frac{\partial g_{\mu\alpha}}{\partial x_\nu}+\frac{\partial g_{\nu\alpha}}{\partial x_\mu}-\frac{\partial g_{\mu\nu}}{\partial x_\alpha}\right)g^{\mu\nu}A_\tau.$$

In accordance with (31) and (29), the last term of this expression may be written

$$\frac{1}{2}\frac{\partial g^{\tau\nu}}{\partial x_\nu}A_\tau+\frac{1}{2}\frac{\partial g^{\tau\mu}}{\partial x_\mu}A_\tau+\frac{1}{\sqrt{-g}}\frac{\partial\sqrt{-g}}{\partial x_\alpha}g^{\mu\nu}A_\tau.$$

As the symbols of the indices of summation are immaterial, the first two terms of this expression cancel the second of the one above. If we then write $g^{\mu\nu}A_\mu=A^\nu$, so that A^ν like A_μ is an arbitrary vector, we finally obtain

$$\Phi = \frac{1}{\sqrt{-g}} \frac{\partial}{\partial x_\nu}\left(\sqrt{-g}\, A^\nu\right) \tag{35}$$

This scalar is the *divergence* of the contravariant vector A^ν.

The "Curl" of a Covariant Vector.—The second term in (26) is symmetrical in the indices μ and ν. Therefore $A_{\mu\nu} - A_{\nu\mu}$ is a particularly simply constructed antisymmetrical tensor. We obtain

$$B_{\mu\nu} = \frac{\partial A_\mu}{\partial x_\nu} - \frac{\partial A_\nu}{\partial x_\mu} \tag{36}$$

Antisymmetrical Extension of a Six-vector.—Applying (27) to an antisymmetrical tensor of the second rank $A_{\mu\nu}$, forming in addition the two equations which arise through cyclic permutations of the indices, and adding these three equations, we obtain the tensor of the third rank

$$B_{\mu\nu\sigma} = A_{\mu\nu\sigma} + A_{\nu\sigma\mu} + A_{\sigma\mu\nu} = \frac{\partial A_{\mu\nu}}{\partial x_\sigma} + \frac{\partial A_{\nu\sigma}}{\partial x_\mu} + \frac{\partial A_{\sigma\mu}}{\partial x_\nu} \tag{37}$$

which it is easy to prove is antisymmetrical.

The Divergence of a Six-vector.—Taking the mixed product of (27) by $g^{\mu\alpha}g^{\nu\beta}$, we also obtain a tensor. The first term on the right-hand side of (27) may be written in the form

$$\frac{\partial}{\partial x_\sigma}\left(g^{\mu\alpha} g^{\nu\beta} A_{\mu\nu}\right) - g^{\mu\alpha}\frac{\partial g^{\nu\beta}}{\partial x_\sigma} A_{\mu\nu} - g^{\nu\beta}\frac{\partial g^{\mu\alpha}}{\partial x_\sigma} A_{\mu\nu}.$$

If we write $A_\sigma^{\alpha\beta}$ for $g^{\mu\alpha}g^{\nu\beta}A_{\mu\nu\sigma}$ and $A^{\alpha\beta}$ for $g^{\mu\alpha}g^{\nu\beta}A_{\mu\nu}$, and in the transformed first term replace

$$\frac{\partial g^{\nu\beta}}{\partial x_\sigma} \text{ and } \frac{\partial g^{\mu\alpha}}{\partial x_\sigma}$$

by their values as given by (34), there results from the right-hand side of (27) an expression consisting of seven terms, of which four cancel, and there remains

$$A_\sigma^{\alpha\beta} = \frac{\partial A^{\alpha\beta}}{\partial x_\sigma} + \{\sigma\gamma, \alpha\} A^{\gamma\beta} + \{\sigma\gamma, \beta\} A^{\alpha\gamma} \tag{38}$$

This is the expression for the extension of a contravariant tensor of the second rank, and corresponding expressions for the extension of contravariant tensors of higher and lower rank may also be formed.

We note that in an analogous way we may also form the extension of a mixed tensor:—

$$A_{\mu\sigma}^{\alpha} = \frac{\partial A_\mu^\alpha}{\partial x_\sigma} - \{\sigma\mu, \tau\} A_\tau^\alpha + \{\sigma\tau, \alpha\} A_\mu^\tau \tag{39}$$

On contracting (38) with respect to the indices β and σ (inner multiplication by δ_β^σ), we obtain the vector

$$A^{\alpha} = \frac{\partial A^{\alpha\beta}}{\partial x_{\beta}} + \{\beta\gamma, \beta\} A^{\alpha\gamma} + \{\beta\gamma, \alpha\} A^{\gamma\beta}.$$

On account of the symmetry of $\{\beta\gamma, \alpha\}$ with respect to the indices β and γ, the third term on the right-hand side vanishes, if $A^{\alpha\beta}$ is, as we will assume, an antisymmetrical tensor. The second term allows itself to be transformed in accordance with (29a). Thus we obtain

$$A^{\alpha} = \frac{1}{\sqrt{-g}} \frac{\partial\left(\sqrt{-g}A^{\alpha\beta}\right)}{\partial x_{\beta}} \qquad (40)$$

This is the expression for the divergence of a contravariant six-vector.

The Divergence of a Mixed Tensor of the Second Rank.—Contracting (39) with respect to the indices α and σ, and taking (29a) into consideration, we obtain

$$\sqrt{-g}A_{\mu} = \frac{\partial\left(\sqrt{-g}A_{\mu}^{\sigma}\right)}{\partial x_{\sigma}} - \{\sigma\mu, \tau\}\sqrt{-g}A_{\tau}^{\sigma} \qquad (41)$$

If we introduce the contravariant tensor $A^{\rho\sigma} = g^{\rho\tau}A_{\tau}^{\sigma}$ in the last term, it assumes the form

$$-[\sigma\mu, \rho]\sqrt{-g}A^{\rho\sigma}.$$

If, further, the tensor $A^{\rho\sigma}$ is symmetrical, this reduces to

$$-\frac{1}{2}\sqrt{-g}\frac{\partial g_{\rho\sigma}}{\partial x_{\mu}}A^{\rho\sigma}.$$

Had we introduced, instead of $A^{\rho\sigma}$, the covariant tensor $A_{\rho\sigma} = g_{\rho\alpha}g_{\sigma\beta}A^{\alpha\beta}$, which is also symmetrical, the last term, by virtue of (31), would assume the form

$$\frac{1}{2}\sqrt{-g}\frac{\partial g^{\rho\sigma}}{\partial x_{\mu}}A_{\rho\sigma}.$$

In the case of symmetry in question, (41) may therefore be replaced by the two forms

$$\sqrt{-g}A_{\mu} = \frac{\partial\left(\sqrt{-g}A_{\mu}^{\sigma}\right)}{\partial x_{\sigma}} - \frac{1}{2}\frac{\partial g_{\rho\sigma}}{\partial x_{\mu}}\sqrt{-g}A^{\rho\sigma} \qquad (41a)$$

$$\sqrt{-g}A_{\mu} = \frac{\partial\left(\sqrt{-g}A_{\mu}^{\sigma}\right)}{\partial x_{\sigma}} + \frac{1}{2}\frac{\partial g^{\rho\sigma}}{\partial x_{\mu}}\sqrt{-g}A_{\rho\sigma} \qquad (41b)$$

which we have to employ later on.

§ 12. THE RIEMANN-CHRISTOFFEL TENSOR

We now seek the tensor which can be obtained from the fundamental tensor *alone*, by differentiation. At first sight the solution seems obvious. We place the fundamental tensor of the $g_{\mu\nu}$ in (27) instead of any given tensor $A_{\mu\nu}$, and thus have a new tensor, namely, the extension of the fundamental tensor. But we easily convince ourselves that this extension vanishes identically. We reach our goal, however, in the following way. In (27) place

$$A_{\mu\nu} = \frac{\partial A_\mu}{\partial x_\nu} - \{\mu\nu, \rho\} A_\rho,$$

i.e. the extension of the four-vector A_μ. Then (with a somewhat different naming of the indices) we get the tensor of the third rank

$$A_{\mu\sigma\tau} = \frac{\partial^2 A_\mu}{\partial x_\sigma \partial x_\tau} - \{\mu\sigma, \rho\}\frac{\partial A_\rho}{\partial x_\tau} - \{\mu\tau, \rho\}\frac{\partial A_\rho}{\partial x_\sigma} - \{\sigma\tau, \rho\}\frac{\partial A_\mu}{\partial x_\rho}$$
$$+ \left[-\frac{\partial}{\partial x_\tau}\{\mu\sigma, \rho\} + \{\mu\tau, \alpha\}\{\alpha\sigma, \rho\} + \{\sigma\tau, \alpha\}\{\alpha\mu, \rho\}\right] A_\rho.$$

This expression suggests forming the tensor $A_{\mu\sigma\tau} - A_{\mu\tau\sigma}$. For, if we do so, the following terms of the expression for $A_{\mu\sigma\tau}$ cancel those of $A_{\mu\tau\sigma}$, the first, the fourth, and the member corresponding to the last term in square brackets; because all these are symmetrical in σ and τ. The same holds good for the sum of the second and third terms. Thus we obtain

$$A_{\mu\sigma\tau} - A_{\mu\tau\sigma} = B^\rho_{\mu\sigma\tau} A_\rho \qquad (42)$$

where

$$B^\rho_{\mu\sigma\tau} = -\frac{\partial}{\partial x_\tau}\{\mu\sigma, \rho\} + \frac{\partial}{\partial x_\sigma}\{\mu\tau, \rho\} - \{\mu\sigma, \alpha\}\{\alpha\tau, \rho\}$$
$$+ \{\mu\tau, \alpha\}\{\alpha\sigma, \rho\} \qquad (43)$$

The essential feature of the result is that on the right side of (42) the A_ρ occur alone, without their derivatives. From the tensor character of $A_{\mu\sigma\tau} - A_{\mu\tau\sigma}$ in conjunction with the fact that A_ρ is an arbitrary vector, it follows, by reason of § 7, that $B^\rho_{\mu\sigma\tau}$ is a tensor (the Riemann-Christoffel tensor).

The mathematical importance of this tensor is as follows: If the continuum is of such a nature that there is a co-ordinate system with reference to which the $g_{\mu\nu}$ are constants, then all the $B^\rho_{\mu\sigma\tau}$ vanish. If we choose any new system of co-ordinates in place of the original ones, the $g_{\mu\nu}$ referred thereto will not be constants, but in consequence of its tensor nature, the transformed components of $B^\rho_{\mu\sigma\tau}$ will still vanish in the new system. Thus the vanishing of the Riemann tensor is a necessary condition that, by an appropriate choice of the system of reference, the $g_{\mu\nu}$ may be constants. In our problem this corresponds to the case in which,* with a suitable choice of the system of reference, the special theory of relativity holds good for a *finite* region of the continuum.

Contracting (43) with respect to the indices τ and ρ we obtain the covariant tensor of second rank

* The mathematicians have proved that this is also a *sufficient* condition.

where

$$\left.\begin{aligned} G_{\mu\nu} &= B^{\rho}_{\mu\nu\rho} = R_{\mu\nu} + S_{\mu\nu} \\ R_{\mu\nu} &= -\frac{\partial}{\partial x_\alpha}\{\mu\nu, \alpha\} + \{\mu\alpha, \beta\}\{\nu\beta, \alpha\} \\ S_{\mu\nu} &= \frac{\partial^2 \log\sqrt{-g}}{\partial x_\mu \partial x_\nu} - \{\mu\nu, \alpha\}\frac{\partial \log\sqrt{-g}}{\partial x_\alpha} \end{aligned}\right\} \quad (44)$$

Note on the Choice of Co-ordinates.—It has already been observed in § 8, in connexion with equation (18a), that the choice of co-ordinates may with advantage be made so that $\sqrt{-g} = 1$. A glance at the equations obtained in the last two sections shows that by such a choice the laws of formation of tensors undergo an important simplification. This applies particularly to $G_{\mu\nu}$, the tensor just developed, which plays a fundamental part in the theory to be set forth. For this specialization of the choice of co-ordinates brings about the vanishing of $S_{\mu\nu}$, so that the tensor $G_{\mu\nu}$ reduces to $R_{\mu\nu}$.

On this account I shall hereafter give all relations in the simplified form which this specialization of the choice of co-ordinates brings with it. It will then be an easy matter to revert to the *generally* covariant equations, if this seems desirable in a special case.

C. THEORY OF THE GRAVITATIONAL FIELD

§ 13. EQUATIONS OF MOTION OF A MATERIAL POINT IN THE GRAVITATIONAL FIELD. EXPRESSION FOR THE FIELD-COMPONENTS OF GRAVITATION

A freely movable body not subjected to external forces moves, according to the special theory of relativity, in a straight line and uniformly. This is also the case, according to the general theory of relativity, for a part of four-dimensional space in which the system of co-ordinates K_0, may be, and is, so chosen that they have the special constant values given in (4).

If we consider precisely this movement from any chosen system of co-ordinates K_1, the body, observed from K_1, moves, according to the considerations in § 2, in a gravitational field. The law of motion with respect to K_1 results without difficulty from the following consideration. With respect to K_0 the law of motion corresponds to a four-dimensional straight line, i.e. to a geodetic line. Now since the geodetic line is defined independently of the system of reference, its equations will also be the equation of motion of the material point with respect to K_1. If we set

$$\Gamma^{\tau}_{\mu\nu} = -\{\mu\nu, \tau\} \quad (45)$$

the equation of the motion of the point with respect to K_1, becomes

$$\frac{d^2 x_\tau}{ds^2} = \Gamma^\tau_{\mu\nu} \frac{dx_\mu}{ds} \frac{dx_\nu}{ds} \qquad (46)$$

We now make the assumption, which readily suggests itself, that this covariant system of equations also defines the motion of the point in the gravitational field in the case when there is no system of reference K_0, with respect to which the special theory of relativity holds good in a finite region. We have all the more justification for this assumption as (46) contains only *first* derivatives of the $g_{\mu\nu}$, between which even in the special case of the existence of K_0, no relations subsist.*

If the $\Gamma^\tau_{\mu\nu}$ vanish, then the point moves uniformly in a straight line. These quantities therefore condition the deviation of the motion from uniformity. They are the components of the gravitational field.

§ 14. THE FIELD EQUATIONS OF GRAVITATION IN THE ABSENCE OF MATTER

We make a distinction hereafter between "gravitational field" and "matter" in this way, that we denote everything but the gravitational field as "matter." Our use of the word therefore includes not only matter in the ordinary sense, but the electromagnetic field as well.

Our next task is to find the field equations of gravitation in the absence of matter. Here we again apply the method employed in the preceding paragraph in formulating the equations of motion of the material point. A special case in which the required equations must in any case be satisfied is that of the special theory of relativity, in which the $g_{\mu\nu}$, have certain constant values. Let this be the case in a certain finite space in relation to a definite system of co-ordinates K_0. Relatively to this system all the components of the Riemann tensor $B^\rho_{\mu\nu\tau}$, defined in (43), vanish. For the space under consideration they then vanish, also in any other system of co-ordinates.

Thus the required equations of the matter-free gravitational field must in any case be satisfied if all $B^\rho_{\mu\sigma\tau}$ vanish. But this condition goes too far. For it is clear that, e.g., the gravitational field generated by a material point in its environment certainly cannot be "transformed away" by any choice of the system of co-ordinates, i.e. it cannot be transformed to the case of constant $g_{\mu\nu}$.

This prompts us to require for the matter-free gravitational field that the symmetrical tensor $G_{\mu\nu}$, derived from the tensor $B^\rho_{\mu\nu\tau}$, shall vanish. Thus we obtain ten equations for the ten quantities $g_{\mu\nu}$, which are satisfied in the special case of the vanishing of all $B^\rho_{\mu\nu\tau}$. With the choice which we have made of a system of co-ordinates, and taking (44) into consideration, the equations for the matter-free field are

* It is only between the second (and first) derivatives that, by § 12, the relations $B^\rho_{\mu\sigma\tau} = 0$ subsist.

$$\left.\begin{aligned}&\frac{\partial \Gamma^{\alpha}_{\mu\nu}}{\partial x_{\alpha}} + \Gamma^{\alpha}_{\mu\beta}\Gamma^{\beta}_{\nu\alpha} = 0\\ &\sqrt{-g} = 1\end{aligned}\right\} \qquad (47)$$

It must be pointed out that there is only a minimum of arbitrariness in the choice of these equations. For besides $G_{\mu\nu}$ there is no tensor of second rank which is formed from the $g_{\mu\nu}$ and its derivatives, contains no derivations higher than second, and is linear in these derivatives.*

These equations, which proceed, by the method of pure mathematics, from the requirement of the general theory of relativity, give us, in combination with the equations of motion (46), to a first approximation Newton's law of attraction, and to a second approximation the explanation of the motion of the perihelion of the planet Mercury discovered by Leverrier (as it remains after corrections for perturbation have been made). These facts must, in my opinion, be taken as a convincing proof of the correctness of the theory.

§ 15. THE HAMILTONIAN FUNCTION FOR THE GRAVITATIONAL FIELD. LAWS OF MOMENTUM AND ENERGY

To show that the field equations correspond to the laws of momentum and energy, it is most convenient to write them in the following Hamiltonian form:

$$\left.\begin{aligned}&\delta \int H\, d\tau = 0\\ &H = g^{\mu\nu}\Gamma^{\alpha}_{\mu\beta}\Gamma^{\beta}_{\nu\alpha}\\ &\sqrt{-g} = 1\end{aligned}\right\} \qquad (47a)$$

where, on the boundary of the finite four-dimensional region of integration which we have in view, the variations vanish.

We first have to show that the form (47a) is equivalent to the equations (47). For this purpose we regard H as a function of the $g^{\mu\nu}$ and the $g^{\mu\nu}_{\sigma}\left(= \partial g^{\mu\nu} / \partial x_{\sigma}\right)$.

Then in the first place

$$\begin{aligned}\delta H &= \Gamma^{\alpha}_{\mu\beta}\Gamma^{\beta}_{\nu\alpha}\delta g^{\mu\nu} + 2g^{\mu\nu}\Gamma^{\alpha}_{\mu\beta}\delta\Gamma^{\beta}_{\nu\alpha}\\ &= -\Gamma^{\alpha}_{\mu\beta}\Gamma^{\beta}_{\nu\alpha}\delta g^{\mu\nu} + 2\Gamma^{\alpha}_{\mu\beta}\delta\left(g^{\mu\nu}\Gamma^{\beta}_{\nu\alpha}\right).\end{aligned}$$

But

$$\delta\left(g^{\mu\nu}\Gamma^{\beta}_{\nu\alpha}\right) = -\frac{1}{2}\delta\left[g^{\mu\nu}g^{\beta\lambda}\left(\frac{\partial g_{\nu\lambda}}{\partial x_{\alpha}} + \frac{\partial g_{\alpha\lambda}}{\partial x_{\nu}} - \frac{\partial g_{\alpha\nu}}{\partial x_{\lambda}}\right)\right].$$

The terms arising from the last two terms in round brackets are of different sign, and result from each other (since the denomination of the summation indices is immaterial)

* Properly speaking, this can be affirmed only of the tensor

$$G_{\mu\nu} + \lambda g_{\mu\nu} g^{\alpha\beta} G_{\alpha\beta}$$

where λ is a constant. If, however, we set this tensor = 0, we come back again to the equations $G_{\mu\nu} = 0$.

through interchange of the indices μ and β. They cancel each other in the expression for δH, because they are multiplied by the quantity $\Gamma^{\alpha}_{\mu\beta}$, which is symmetrical with respect to the indices μ and β. Thus there remains only the first term in round brackets to be considered, so that, taking (31) into account, we obtain

$$\delta H = -\Gamma^{\alpha}_{\mu\beta}\Gamma^{\beta}_{\nu\alpha}\delta g^{\mu\nu} + \Gamma^{\alpha}_{\mu\beta}\delta g^{\mu\beta}_{\alpha}$$

Thus

$$\left.\begin{aligned} \frac{\partial H}{\partial g^{\mu\nu}} &= -\Gamma^{\alpha}_{\mu\beta}\Gamma^{\beta}_{\nu\alpha} \\ \frac{\partial H}{\partial g^{\mu\nu}_{\sigma}} &= \Gamma^{\sigma}_{\mu\nu} \end{aligned}\right\} \tag{48}$$

Carrying out the variation in (47a), we get in the first place

$$\frac{\partial}{\partial x_{\alpha}}\left(\frac{\partial H}{\partial g^{\mu\nu}_{\alpha}}\right) - \frac{\partial H}{\partial g^{\mu\nu}} = 0, \tag{47b}$$

which, on account of (48), agrees with (47), as was to be proved.

If we multiply (47b) by $g^{\mu\nu}_{\sigma}$, then because

$$\frac{\partial g^{\mu\nu}_{\sigma}}{\partial x_{\alpha}} = \frac{\partial g^{\mu\nu}_{\alpha}}{\partial x_{\sigma}}$$

and, consequently,

$$g^{\mu\nu}_{\sigma}\frac{\partial}{\partial x_{\alpha}}\left(\frac{\partial H}{\partial g^{\mu\nu}_{\alpha}}\right) = \frac{\partial}{\partial x_{\alpha}}\left(g^{\mu\nu}_{\sigma}\frac{\partial H}{\partial g^{\mu\nu}_{\alpha}}\right) - \frac{\partial H}{\partial g^{\mu\nu}_{\alpha}}\frac{\partial g^{\mu\nu}_{\alpha}}{\partial x_{\sigma}},$$

we obtain the equation

$$\frac{\partial}{\partial x_{\alpha}}\left(g^{\mu\nu}_{\sigma}\frac{\partial H}{\partial g^{\mu\nu}_{\alpha}}\right) - \frac{\partial H}{\partial x_{\sigma}} = 0$$

or*

$$\left.\begin{aligned} &\frac{\partial t^{\alpha}_{\sigma}}{\partial x_{\alpha}} = 0 \\ &-2\kappa t^{\alpha}_{\sigma} = g^{\mu\nu}_{\sigma}\frac{\partial H}{\partial g^{\mu\nu}_{\alpha}} - \delta^{\alpha}_{\sigma}H \end{aligned}\right\} \tag{49}$$

where, on account of (48), the second equation of (47), and (34)

$$\kappa t^{\alpha}_{\sigma} = \frac{1}{2}\delta^{\alpha}_{\sigma}g^{\mu\nu}\Gamma^{\lambda}_{\mu\beta}\Gamma^{\beta}_{\nu\lambda} - g^{\mu\nu}\Gamma^{\alpha}_{\mu\beta}\Gamma^{\beta}_{\nu\sigma} \tag{50}$$

It is to be noticed that t^{α}_{σ} is not a tensor; on the other hand (49) applies to all systems of co-ordinates for which $\sqrt{-g} = 1$. This equation expresses the law of conservation of momentum and of energy for the gravitational field. Actually the integration of this equation over a three-dimensional volume V yields the four equations

* The reason for the introduction of the factor -2κ will be apparent later.

$$\frac{d}{dx_4}\int t_\sigma^4 dV = \int \left(l t_\sigma^1 + m t_\sigma^2 + n t_\sigma^3\right) dS \qquad (49a)$$

where *l*, *m*, *n* denote the direction-cosines of direction of the inward drawn normal at the element *dS* of the bounding surface (in the sense of Euclidean geometry). We recognize in this the expression of the laws of conservation in their usual form. The quantities t_σ^α we call the "energy components" of the gravitational field.

I will now give equations (47) in a third form, which is particularly useful for a vivid grasp of our subject. By multiplication of the field equations (47) by $g^{\nu\sigma}$ these are obtained in the "mixed" form. Note that

$$g^{\nu\sigma}\frac{\partial\Gamma_{\mu\nu}^\alpha}{\partial x_\alpha} = \frac{\partial}{\partial x_\alpha}\left(g^{\nu\sigma}\Gamma_{\mu\nu}^\alpha\right) - \frac{\partial g^{\nu\sigma}}{\partial x_\alpha}\Gamma_{\mu\nu}^\alpha$$

which quantity, by reason of (34), is equal to

$$\frac{\partial}{\partial x_\alpha}\left(g^{\nu\sigma}\Gamma_{\mu\nu}^\alpha\right) - g^{\nu\beta}\Gamma_{\alpha\beta}^\sigma\Gamma_{\mu\nu}^\alpha - g^{\sigma\beta}\Gamma_{\beta\alpha}^\nu\Gamma_{\mu\nu}^\alpha,$$

or (with different symbols for the summation indices)

$$\frac{\partial}{\partial x_\alpha}\left(g^{\sigma\beta}\Gamma_{\mu\beta}^\alpha\right) - g^{\gamma\delta}\Gamma_{\gamma\beta}^\sigma\Gamma_{\delta\mu}^\beta - g^{\nu\sigma}\Gamma_{\mu\beta}^\alpha\Gamma_{\nu\alpha}^\beta.$$

The third term of this expression cancels with the one arising from the second term of the field equations (47); using relation (50), the second term may be written

$$\kappa\left(t_\mu^\sigma - \frac{1}{2}\delta_\mu^\sigma t\right)$$

where $t = t_\alpha^\alpha$. Thus instead of equations (47) we obtain

$$\left.\begin{aligned}\frac{\partial}{\partial x_\alpha}\left(g^{\sigma\beta}\Gamma_{\mu\beta}^\alpha\right) &= -\kappa\left(t_\mu^\sigma - \tfrac{1}{2}\delta_\mu^\sigma t\right)\\ \sqrt{-g} &= 1\end{aligned}\right\} \qquad (51)$$

§ 16. THE GENERAL FORM OF THE FIELD EQUATIONS OF GRAVITATION

The field equations for matter-free space formulated in § 15 are to be compared with the field equation

$$\nabla^2\phi = 0$$

of Newton's theory. We require the equation corresponding to Poisson's equation

$$\nabla^2\phi = 4\pi\kappa\rho,$$

where ρ denotes the density of matter.

The special theory of relativity has led to the conclusion that inert mass is nothing more or less than energy, which finds its complete mathematical expression in a symmetrical tensor of second rank, the energy-tensor. Thus in the general theory of relativity we must introduce a corresponding energy-tensor of matter T_σ^α , which, like the energy-components t_σ [equations (49) and (50)] of the gravitational field, will have mixed character, but will pertain to a symmetrical covariant tensor.*

The system of equation (51) shows how this energy-tensor (corresponding to the density ρ in Poisson's equation) is to be introduced into the field equations of gravitation. For if we consider a complete system (e.g. the solar system), the total mass of the system, and therefore its total gravitating action as well, will depend on the total energy of the system, and therefore on the ponderable energy together with the gravitational energy. This will allow itself to be expressed by introducing into (51), in place of the energy-components of the gravitational field alone, the sums $t_\mu^\sigma + T_\mu^\sigma$ of the energy-components of matter and of gravitational field. Thus instead of (51) we obtain the tensor equation

$$\left.\begin{aligned} \frac{\partial}{\partial x_\alpha}\left(g^{\sigma\beta}\Gamma_{\mu\beta}^\alpha\right) &= -\kappa\left[\left(t_\mu^\sigma + T_\mu^\sigma\right) - \frac{1}{2}\delta_\mu^\sigma(t+T)\right] \\ \sqrt{-g} &= 1 \end{aligned}\right\} \qquad (52)$$

where we have set $T = T_\mu^\mu$ (Laue's scalar). These are the required general field equations of gravitation in mixed form. Working back from these, we have in place of (47)

$$\left.\begin{aligned} \frac{\partial}{\partial x_\alpha}\Gamma_{\mu\nu}^\alpha + \Gamma_{\mu\beta}^\alpha\Gamma_{\mu\nu}^\beta &= -\kappa\left(T_{\mu\nu} - \frac{1}{2}g_{\mu\nu}T\right), \\ \sqrt{-g} &= 1 \end{aligned}\right\} \qquad (53)$$

It must be admitted that this introduction of the energy-tensor of matter is not justified by the relativity postulate alone. For this reason we have here deduced it from the requirement that the energy of the gravitational field shall act gravitatively in the same way as any other kind of energy. But the strongest reason for the choice of these equations lies in their consequence, that the equations of conservation of momentum and energy, corresponding exactly to equations (49) and (49a), hold good for the components of the total energy. This will be shown in § 17.

§ 17. The Laws of Conservation in the General Case

Equation (52) may readily be transformed so that the second term on the right-hand side vanishes. Contract (52) with respect to the indices μ and σ, and after multiplying the resulting equation by $\frac{1}{2}\delta_\mu^\sigma$, subtract it from equation (52). This gives

* $g_{\alpha\tau}T_\sigma^\alpha = T_{\sigma\tau}$ and $g^{\sigma\beta}T_\sigma^\alpha = T^{\alpha\beta}$ are to be symmetrical tensors.

$$\frac{\partial}{\partial x_\alpha}\left(g^{\sigma\beta}\Gamma^\alpha_{\mu\beta} - \frac{1}{2}\delta^\sigma_\mu g^{\lambda\beta}\Gamma^\alpha_{\lambda\beta}\right) = -\kappa\left(t^\sigma_\mu + T^\sigma_\mu\right) \qquad (52a)$$

On this equation we perform the operation $\partial/\partial x_\sigma$. We have

$$\frac{\partial^2}{\partial x_\alpha \partial x_\sigma}\left(g^{\sigma\beta}\Gamma^\alpha_{\beta\mu}\right) = -\frac{1}{2}\frac{\partial^2}{\partial x_\alpha \partial x_\sigma}\left[g^{\sigma\beta}g^{\alpha\lambda}\left(\frac{\partial g_{\mu\lambda}}{\partial x_\beta} + \frac{\partial g_{\beta\lambda}}{\partial x_\mu} - \frac{\partial g_{\mu\beta}}{\partial x_\lambda}\right)\right].$$

The first and third terms of the round brackets yield contributions which cancel one another, as may be seen by interchanging, in the contribution of the third term, the summation indices α and σ on the one hand, and β and λ on the other. The second term may be re-modelled by (31), so that we have

$$\frac{\partial^2}{\partial x_\alpha \partial x_\sigma}\left(g^{\sigma\beta}\Gamma^\alpha_{\mu\beta}\right) = \frac{1}{2}\frac{\partial^3 g^{\alpha\beta}}{\partial x_\alpha \partial x_\beta \partial x_\mu} \qquad (54)$$

The second term on the left-hand side of (52a) yields in the first place

$$-\frac{1}{2}\frac{\partial^2}{\partial x_\alpha \partial x_\mu}\left(g^{\lambda\beta}\Gamma^\alpha_{\lambda\beta}\right)$$

or

$$\frac{1}{4}\frac{\partial^2}{\partial x_\alpha \partial x_\mu}\left[g^{\lambda\beta}g^{\alpha\delta}\left(\frac{\partial g_{\delta\lambda}}{\partial x_\beta} + \frac{\partial g_{\delta\beta}}{\partial x_\lambda} - \frac{\partial g_{\lambda\beta}}{\partial x_\delta}\right)\right].$$

With the choice of co-ordinates which we have made, the term deriving from the last term in round brackets disappears by reason of (29). The other two may be combined, and together, by (31), they give

$$-\frac{1}{2}\frac{\partial^3 g^{\alpha\beta}}{\partial x_\alpha \partial x_\beta \partial x_\mu},$$

so that in consideration of (54), we have the identity

$$\frac{\partial^2}{\partial x_\alpha \partial x_\sigma}\left(g^{\sigma\beta}\Gamma^\alpha_{\mu\beta} - \frac{1}{2}\delta^\sigma_\mu g^{\lambda\beta}\Gamma^\alpha_{\lambda\beta}\right) \equiv 0 \qquad (55)$$

From (55) and (52a), it follows that

$$\frac{\partial\left(t^\sigma_\mu + T^\sigma_\mu\right)}{\partial x_\sigma} = 0 \qquad (56)$$

Thus it results from our field equations of gravitation that the laws of conservation of momentum and energy are satisfied. This may be seen most easily from the consideration which leads to equation (49a); except that here, instead of the energy components t^σ of the gravitational field, we have to introduce the totality of the energy components of matter and gravitational field.

§ 18. The Laws of Momentum and Energy for Matter, as a Consequence of the Field Equations

Multiplying (53) by $\partial g^{\mu\nu}/\partial x_\sigma$, we obtain, by the method adopted in § 15, in view of the vanishing of

$$g_{\mu\nu}\frac{\partial g^{\mu\nu}}{\partial x_\sigma},$$

the equation

$$\frac{\partial t_\sigma^\alpha}{\partial x_\alpha}+\frac{1}{2}\frac{\partial g^{\mu\nu}}{\partial x_\sigma}T_{\mu\nu}=0,$$

or, in view of (56),

$$\frac{\partial T_\sigma^\alpha}{\partial x_\alpha}+\frac{1}{2}\frac{\partial g^{\mu\nu}}{\partial x_\sigma}T_{\mu\nu}=0 \qquad (57)$$

Comparison with (41b) shows that with the choice of system of co-ordinates which we have made, this equation predicates nothing more or less than the vanishing of divergence of the material energy-tensor. Physically, the occurrence of the second term on the left-hand side shows that laws of conservation of momentum and energy do not apply in the strict sense for matter alone, or else that they apply only when the $g^{\mu\nu}$ are constant, i.e. when the field intensities of gravitation vanish. This second term is an expression for momentum, and for energy, as transferred per unit of volume and time from the gravitational field to matter. This is brought out still more clearly by re-writing (57) in the sense of (41) as

$$\frac{\partial T_\sigma^\alpha}{\partial x_\alpha}=-\Gamma_{\alpha\sigma}^\beta T_\beta^\alpha \qquad (57a)$$

The right side expresses the energetic effect of the gravitational field on matter.

Thus the field equations of gravitation contain four conditions which govern the course of material phenomena. They give the equations of material phenomena completely, if the latter is capable of being characterized by four differential equations independent of one another.*

D. MATERIAL PHENOMENA

The mathematical aids developed in part B enable us forthwith to generalize the physical laws of matter (hydrodynamics, Maxwell's electrodynamics), as they are formulated in the special theory of relativity, so that they will fit in with the general theory of relativity. When this is done, the general principle of relativity does not indeed afford us a further limitation of possibilities; but it makes us acquainted with the influence of the gravitational field on all processes, without our having to introduce any new hypothesis whatever.

* On this question cf. H. Hilbert, Nachr. d. K. Gesellsch. d. Wiss. zu Göttingen, Math.-phys. Klasse, 1915, p. 3.

Hence it comes about that it is not necessary to introduce definite assumptions as to the physical nature of matter (in the narrower sense). In particular it may remain an open question whether the theory of the electromagnetic field in conjunction with that of the gravitational field furnishes a sufficient basis for the theory of matter or not. The general postulate of relativity is unable on principle to tell us anything about this. It must remain to be seen, during the working out of the theory, whether electromagnetics and the doctrine of gravitation are able in collaboration to perform what the former by itself is unable to do.

§ 19. EULER'S EQUATIONS FOR A FRICTIONLESS ADIABATIC FLUID

Let p and ρ be two scalars, the former of which we call the "pressure," the latter the "density" of a fluid; and let an equation subsist between them. Let the contravariant symmetrical tensor

$$T^{\alpha\beta} = -g^{\alpha\beta} p + \rho \frac{dx_\alpha}{ds} \frac{dx_\beta}{ds} \tag{58}$$

be the contravariant energy-tensor of the fluid. To it belongs the covariant tensor

$$T_{\mu\nu} = -g_{\mu\nu} p + g_{\mu\alpha} g_{\mu\beta} \frac{dx_\alpha}{ds} \frac{dx_\beta}{ds} \rho, \tag{58a}$$

as well as the mixed tensor*

$$T_\sigma^\alpha = -\delta_\sigma^\alpha p + g_{\sigma\beta} \frac{dx_\beta}{ds} \frac{dx_\alpha}{ds} \rho \tag{58b}$$

Inserting the right-hand side of (58b) in (57a), we obtain the Eulerian hydrodynamical equations of the general theory of relativity. They give, in theory, a complete solution of the problem of motion, since the four equations (57a), together with the given equation between p and ρ, and the equation

$$g_{\alpha\beta} \frac{dx_\alpha}{ds} \frac{dx_\beta}{ds} = 1,$$

are sufficient, $g_{\alpha\beta}$ being given, to define the six unknowns

$$p, \rho, \frac{dx_1}{ds}, \frac{dx_2}{ds}, \frac{dx_3}{ds}, \frac{dx_4}{ds}.$$

If the $g_{\mu\nu}$ are also unknown, the equations (53) are brought in. These are eleven equations for defining the ten functions $g_{\mu\nu}$, so that these functions appear over-defined. We must remember, however, that the equations (57a) are already contained in the equations (53), so that the latter represent only seven independent equations. There is good reason for this lack of definition, in that the wide freedom of the choice of coordinates

* For an observer using a system of reference in the sense of the special theory of relativity for an infinitely small region, and moving with it, the density of energy T_4^4 equals $\rho - p$. This gives the definition of ρ. Thus ρ is not constant for an incompressible fluid.

causes the problem to remain mathematically undefined to such a degree that three of the functions of space may be chosen at will.*

§ 20. MAXWELL'S ELECTROMAGNETIC FIELD EQUATIONS FOR FREE SPACE

Let ϕ_ν be the components of a covariant vector—the electromagnetic potential vector. From them we form, in accordance with (36), the components $F_{\rho\sigma}$ of the covariant six-vector of the electromagnetic field, in accordance with the system of equations

$$F_{\rho\sigma} = \frac{\partial \phi_\rho}{\partial x_\sigma} - \frac{\partial \phi_\sigma}{\partial x_\rho} \tag{59}$$

It follows from (59) that the system of equations

$$\frac{\partial F_{\rho\sigma}}{\partial x_\tau} + \frac{\partial F_{\sigma\tau}}{\partial x_\rho} + \frac{\partial F_{\tau\rho}}{\partial x_\sigma} = 0 \tag{60}$$

is satisfied, its left side being, by (37), an antisymmetrical tensor of the third rank. System (60) thus contains essentially four equations which are written out as follows:—

$$\left.\begin{aligned} \frac{\partial F_{23}}{\partial x_4} + \frac{\partial F_{34}}{\partial x_2} + \frac{\partial F_{42}}{\partial x_3} = 0 \\ \frac{\partial F_{34}}{\partial x_1} + \frac{\partial F_{41}}{\partial x_3} + \frac{\partial F_{13}}{\partial x_4} = 0 \\ \frac{\partial F_{41}}{\partial x_2} + \frac{\partial F_{12}}{\partial x_4} + \frac{\partial F_{24}}{\partial x_1} = 0 \\ \frac{\partial F_{12}}{\partial x_3} + \frac{\partial F_{23}}{\partial x_1} + \frac{\partial F_{31}}{\partial x_2} = 0 \end{aligned}\right\} \tag{60a}$$

This system corresponds to the second of Maxwell's systems of equations. We recognize this at once by setting

$$\left.\begin{aligned} F_{23} = H_x, \quad F_{14} = E_x \\ F_{31} = H_y, \quad F_{24} = E_y \\ F_{12} = H_z, \quad F_{34} = E_z \end{aligned}\right\} \tag{61}$$

Then in place of (60a) we may set, in the usual notation of three-dimensional vector analysis,

$$\left.\begin{aligned} -\frac{\partial H}{\partial t} = \text{curl } E \\ \text{div } H = 0 \end{aligned}\right\} \tag{60b}$$

We obtain Maxwell's first system by generalizing the form given by Minkowski. We introduce the contravariant six-vector associated with $F^{\alpha\beta}$

* On the abandonment of the choice of co-ordinates with g = - 1, there remain four functions of space with liberty of choice, corresponding to the four arbitrary functions at our disposal in the choice of co-ordinates.

$$F^{\mu\nu} = g^{\mu\alpha} g^{\nu\beta} F_{\alpha\beta} \tag{62}$$

and also the contravariant vector J^{μ} of the density of the electric current. Then, taking (40) into consideration, the following equations will be invariant for any substitution whose invariant is unity (in agreement with the chosen co-ordinates):—

$$\frac{\partial}{\partial x_{\nu}} F^{\mu\nu} = J^{\mu} \tag{63}$$

Let

$$\left.\begin{array}{ll} F^{23} = H'_x, & F^{14} = -E'_x \\ F^{31} = H'_y, & F^{24} = -E'_y \\ F^{12} = H'_z, & F^{34} = -E'_z \end{array}\right\} \tag{64}$$

which quantities are equal to the quantities $H_x \ldots E^z$ in the special case of the restricted theory of relativity; and in addition

$$J^1 = j_x,\ J^2 = j_y,\ J^3 = j_z,\ J^4 = \rho,$$

we obtain in place of (63)

$$\left.\begin{array}{l} \dfrac{\partial E'}{\partial t} + j = \text{curl } H' \\ \text{div } E' = \rho \end{array}\right\} \tag{63a}$$

The equations (60), (62), and (63) thus form the generalization of Maxwell's field equations for free space, with the convention which we have established with respect to the choice of co-ordinates.

The Energy-components of the Electromagnetic Field.—We form the inner product

$$\kappa_{\sigma} = F_{\sigma\mu} J^{\mu} \tag{65}$$

By (61) its components, written in the three-dimensional manner, are

$$\left.\begin{array}{l} \kappa_1 = \rho E_x + [j \cdot H]^x \\ \vdots \quad \vdots \quad \vdots \quad \vdots \\ \kappa_4 = -(jE) \end{array}\right\} \tag{65a}$$

κ_{σ} is a covariant vector the components of which are equal to the negative momentum, or, respectively, the energy, which is transferred from the electric masses to the electromagnetic field per unit of time and volume. If the electric masses are free, that is, under the sole influence of the electromagnetic field, the covariant vector κ_{σ} will vanish.

To obtain the energy-components T_σ^ν of the electromagnetic field, we need only give to equation $\kappa_\sigma = 0$ the form of equation (57). From (63) and (65) we have in the first place

$$\kappa_\sigma = F_{\sigma\mu}\frac{\partial F^{\mu\nu}}{\partial x_\nu} = \frac{\partial}{\partial x_\nu}\left(F_{\sigma\mu}F^{\mu\nu}\right) - F^{\mu\nu}\frac{\partial F_{\sigma\mu}}{\partial x_\nu}.$$

The second term of the right-hand side, by reason of (60), permits the transformation

$$F^{\mu\nu}\frac{\partial F_{\sigma\mu}}{\partial x_\nu} - \frac{1}{2}F^{\mu\nu}\frac{\partial F_{\mu\nu}}{\partial x_\sigma} = -\frac{1}{2}g^{\mu\alpha}g^{\nu\beta}F_{\alpha\beta}\frac{\partial F_{\mu\nu}}{\partial x_\sigma},$$

which latter expression may, for reasons of symmetry, also be written in the form

$$-\frac{1}{4}\left[g^{\mu\alpha}g^{\nu\beta}F_{\alpha\beta}\frac{\partial F_{\mu\nu}}{\partial x_\sigma} + g^{\mu\alpha}g^{\nu\beta}\frac{\partial F_{\alpha\beta}}{\partial x_\sigma}F_{\mu\nu}\right].$$

But for this we may set

$$-\frac{1}{4}\frac{\partial}{\partial x_\sigma}\left(g^{\mu\alpha}g^{\nu\beta}F_{\alpha\beta}F_{\mu\nu}\right) + \frac{1}{4}F_{\alpha\beta}F_{\mu\nu}\frac{\partial}{\partial x_\sigma}\left(g^{\mu\alpha}g^{\nu\beta}\right).$$

The first of these terms is written more briefly

$$-\frac{1}{4}\frac{\partial}{\partial x_\sigma}\left(F^{\mu\nu}F_{\mu\nu}\right);$$

the second, after the differentiation is carried out, and after some reduction, results in

$$-\frac{1}{2}F^{\mu\tau}F_{\mu\nu}g^{\nu\rho}\frac{\partial g_{\sigma\tau}}{\partial x_\sigma}.$$

Taking all three terms together we obtain the relation

$$\kappa_\sigma = \frac{\partial T_\sigma^\nu}{\partial x_\nu} - \frac{1}{2}g^{\tau\mu}\frac{\partial g_{\mu\nu}}{\partial x_\sigma}T_\tau^\nu \tag{66}$$

where

$$T_\sigma^\nu = -F_{\sigma\alpha}F^{\nu\alpha} + \frac{1}{4}\delta_\sigma^\nu F_{\alpha\beta}F^{\alpha\beta}.$$

Equation (66), if κ_σ vanishes, is, on account of (30), equivalent to (57) or (57a) respectively. Therefore the T_σ^ν are the energy-components of the electromagnetic field. With the help of (61) and (64), it is easy to show that these energy-components of the electromagnetic field in the case of the special theory of relativity give the well-known Maxwell-Poynting expressions.

We have now deduced the general laws which are satisfied by the gravitational field and matter, by consistently using a system of co-ordinates for which $\sqrt{-g} = 1$. We have thereby achieved a considerable simplification of formulæ and calculations, without failing to comply with the requirement of general covariance; for we have drawn our equations from generally covariant equations by specializing the system of co-ordinates.

Still the question is not without a formal interest, whether with a correspondingly generalized definition of the energy-components of gravitational field and matter, even without specializing the system of co-ordinates, it is possible to formulate laws of conservation in the form of equation (56), and field equations of gravitation of the same nature as (52) or (52a), in such a manner that on the left we have a divergence (in the ordinary sense), and on the right the sum of the energy-components of matter and gravitation. I have found that in both cases this is actually so. But I do not think that the communication of my somewhat extensive reflexions on this subject would be worth while, because after all they do not give us anything that is materially new.

E

§ 21. NEWTON'S THEORY AS A FIRST APPROXIMATION

As has already been mentioned more than once, the special theory of relativity as a special case of the general theory is characterized by the $g_{\mu\nu}$ having the constant values (4). From what has already been said, this means complete neglect of the effects of gravitation. We arrive at a closer approximation to reality by considering the case where the $g_{\mu\nu}$ differ from the values of (4) by quantities which are small compared with 1, and neglecting small quantities of second and higher order. (First point of view of approximation.)

It is further to be assumed that in the space-time territory under consideration the $g_{\mu\nu}$ at spatial infinity, with a suitable choice of co-ordinates, tend toward the values (4); i.e. we are considering gravitational fields which may be regarded as generated exclusively by matter in the finite region.

It might be thought that these approximations must lead us to Newton's theory. But to that end we still need to approximate the fundamental equations from a second point of view. We give our attention to the motion of a material point in accordance with the equations (16). In the case of the special theory of relativity the components

$$\frac{dx_1}{ds}, \frac{dx_2}{ds}, \frac{dx_3}{ds}$$

may take on any values. This signifies that any velocity

$$v = \sqrt{\left(\frac{dx_1}{dx_4}\right)^2 + \left(\frac{dx_2}{dx_4}\right)^2 + \left(\frac{dx_3}{dx_4}\right)^2}$$

may occur, which is less than the velocity of light *in vacuo*. If we restrict ourselves to the case which almost exclusively offers itself to our experience, of v being small as compared with the velocity of light, this denotes that the components

$$\frac{dx_1}{ds}, \frac{dx_2}{ds}, \frac{dx_3}{ds}$$

are to be treated as small quantities, while dx_4/ds, to the second order of small quantities, is equal to one. (Second point of view of approximation.)

Now we remark that from the first point of view of approximation the magnitudes $\Gamma^{\tau}_{\mu\nu}$ are all small magnitudes of at least the first order. A glance at (46) thus shows that in this equation, from the second point of view of approximation, we have to consider only terms for which $\mu = \nu = 4$. Restricting ourselves to terms of lowest order we first obtain in place of (46) the equations

$$\frac{d^2 x_\tau}{dt^2} = \Gamma^{\tau}_{44}$$

where we have set $ds = dx_4 = dt$; or with restriction to terms which from the first point of view of approximation are of first order:—

$$\frac{d^2 x_\tau}{dt^2} = [44, \tau] \quad (\tau = 1, 2, 3)$$

$$\frac{d^2 x_4}{dt^2} = -[44, 4]$$

If in addition we suppose the gravitational field to be a quasi-static field, by confining ourselves to the case where the motion of the matter generating the gravitational field is but slow (in comparison with the velocity of the propagation of light), we may neglect on the right-hand side differentiations with respect to the time in comparison with those with respect to the space co-ordinates, so that we have

$$\frac{d^2 x_\tau}{dt^2} = -\frac{1}{2}\frac{\partial g_{44}}{\partial x_\tau} \quad (\tau = 1, 2, 3) \tag{67}$$

This is the equation of motion of the material point according to Newton's theory, in which $\frac{1}{2} g_{44}$ plays the part of the gravitational potential. What is remarkable in this result is that the component g_{44} of the fundamental tensor alone defines, to a first approximation, the motion of the material point.

We now turn to the field equations (53). Here we have to take into consideration that the energy-tensor of "matter" is almost exclusively defined by the density of matter in the narrower sense, i.e. by the second term of the right-hand side of (58) [or, respectively, (58a) or (58b)]. If we form the approximation in question, all the components vanish with the one exception of $T_{44} = \rho = T$. On the left-hand side of (53) the second term is a small quantity of second order; the first yields, to the approximation in question,

$$\frac{\partial}{\partial x_1}[\mu\nu, 1] + \frac{\partial}{\partial x_2}[\mu\nu, 2] + \frac{\partial}{\partial x_3}[\mu\nu, 3] - \frac{\partial}{\partial x_4}[\mu\nu, 4].$$

For $\mu = \nu = 4$, this gives, with the omission of terms differentiated with respect to time,

$$-\frac{1}{2}\left(\frac{\partial^2 g_{44}}{\partial x_1^2} + \frac{\partial^2 g_{44}}{\partial x_2^2} + \frac{\partial^2 g_{44}}{\partial x_3^2}\right) = -\frac{1}{2}\nabla^2 g_{44}.$$

The last of equations (53) thus yields

$$\nabla^2 g_{44} = \kappa\rho \tag{68}$$

The equations (67) and (68) together are equivalent to Newton's law of gravitation.

By (67) and (68) the expression for the gravitational potential becomes

$$-\frac{\kappa}{8\pi}\int\frac{\rho d\tau}{r} \tag{68a}$$

while Newton's theory, with the unit of time which we have chosen, gives

$$-\frac{K}{c^2}\int\frac{\rho d\tau}{r}$$

in which K denotes the constant $6 \cdot 7 \times 10^{-8}$, usually called the constant of gravitation. By comparison we obtain

$$\kappa = \frac{8\pi K}{c^2} = 1 \cdot 87 \times 10^{-27} \tag{69}$$

§ 22. BEHAVIOUR OF RODS AND CLOCKS IN THE STATIC GRAVITATIONAL FIELD. BENDING OF LIGHT-RAYS. MOTION OF THE PERIHELION OF A PLANETARY ORBIT

To arrive at Newton's theory as a first approximation we had to calculate only one component, g_{44}, of the ten $g_{\mu\nu}$ of the gravitational field, since this component alone enters into the first approximation, (67), of the equation for the motion of the material point in the gravitational field. From this, however, it is already apparent that other components of the $g_{\mu\nu}$ must differ from the values given in (4) by small quantities of the first order. This is required by the condition $g = -1$.

For a field-producing point mass at the origin of co-ordinates, we obtain, to the first approximation, the radially symmetrical solution

$$\left.\begin{aligned} g_{\rho\sigma} &= -\delta_{\rho\sigma} - \alpha\frac{x_\rho x_\sigma}{r^3} \quad (\rho, \sigma = 1, 2, 3) \\ g_{\rho 4} &= g_{4\rho} = 0 \qquad (\rho = 1, 2, 3) \\ g_{44} &= 1 - \frac{\alpha}{r} \end{aligned}\right\} \tag{70}$$

where $\delta_{\rho\sigma}$ is 1 or 0, respectively, accordingly as $\rho = \sigma$ or $\rho \neq \sigma$, and r is the quantity $+\sqrt{x_1^2 + x_2^2 + x_3^2}$ on account of (68a)

$$\alpha = \frac{\kappa M}{4\pi}, \tag{70a}$$

if M denotes the field-producing mass. It is easy to verify that the field equations (outside the mass) are satisfied to the first order of small quantities.

We now examine the influence exerted by the field of the mass M upon the metrical properties of space. The relation

$$ds^2 = g_{\mu\nu} dx_\mu dx_\nu.$$

always holds between the "locally" (§ 4) measured lengths and times ds on the one hand, and the differences of co-ordinates dx_ν on the other hand.

For a unit-measure of length laid "parallel" to the axis of x, for example, we should have to set $ds^2 = -1$; $dx_2 = dx_3 = dx_4 = 0$. Therefore $-1 = g_{11} dx_1^2$. If, in addition, the unit-measure lies on the axis of x, the first of equations (70) gives

$$g_{11} = -\left(1 + \frac{\alpha}{r}\right).$$

From these two relations it follows that, correct to a first order of small quantities,

$$dx = 1 - \frac{\alpha}{2r} \qquad (71)$$

The unit measuring-rod thus appears a little shortened in relation to the system of co-ordinates by the presence of the gravitational field, if the rod is laid along a radius.

In an analogous manner we obtain the length of co-ordinates in tangential direction if, for example, we set

$$ds^2 = -1;\ dx_1 = dx_3 = dx_4 = 0;\ x_1 = r,\ x_2 = x_3 = 0.$$

The result is

$$-1 = g_{22} dx_2^2 = -dx_2^2 \qquad (71a)$$

With the tangential position, therefore, the gravitational field of the point of mass has no influence on the length of rod.

Thus Euclidean geometry does not hold even to a first approximation in the gravitational field, if we wish to take one and the same rod, independently of its place and orientation, as a realization of the same interval; although, to be sure, a glance at (70a) and (69) shows that the deviations to be expected are much too slight to be noticeable in measurements of the earth's surface.

Further, let us examine the rate of a unit clock, which is arranged to be at rest in a static gravitational field. Here we have for a clock period $ds = 1$; $dx_1 = dx_2 = dx_3 = 0$ Therefore

$$1 = g_{44} dx_4^2;$$

$$dx_4 = \frac{1}{\sqrt{g_{44}}} = \frac{1}{\sqrt{(1 + (g_{44} - 1))}} = 1 - \frac{1}{2}(g_{44} - 1)$$

or

$$dx_4 = 1 + \frac{\kappa}{8\pi} \int \rho \frac{d\tau}{r} \qquad (72)$$

Thus the clock goes more slowly if set up in the neighbourhood of ponderable masses.

From this it follows that the spectral lines of light reaching us from the surface of large stars must appear displaced towards the red and of the spectrum.*

We now examine the course of light-rays in the static gravitational field. By the special theory of relativity the velocity of light is given by the equation

$$-dx_1^2 - dx_2 - dx_3^2 + dx_4^2 = 0$$

and therefore by the general theory of relativity by the equation

$$ds^2 = g_{\mu\nu} dx_\mu dx_\nu = 0 \qquad (73)$$

If the direction, i.e. the ratio dx_1: dx_2: dx_3 is given, equation (73) gives the quantities

$$\frac{dx_1}{dx_4}, \frac{dx_2}{dx_4}, \frac{dx_3}{dx_4}$$

and accordingly the velocity

$$\sqrt{\left(\frac{dx_1}{dx_4}\right)^2 + \left(\frac{dx_2}{dx_4}\right)^2 + \left(\frac{dx_3}{dx_4}\right)^2} = \gamma$$

defined in the sense of Euclidean geometry. We easily recognize that the course of the light-rays must be bent with regard to the system of co-ordinates, if the $g_{\mu\nu}$ are not constant. If n is a direction perpendicular to the propagation of light, the Huyghens principle shows that the light-ray, envisaged in the plane (γ, n), has the curvature $-\partial\gamma/\partial n$.

We examine the curvature undergone by a ray of light passing by a mass M at the distance Δ. If we choose the system of co-ordinates in agreement with the accompanying diagram, the total bending of the ray (calculated positively if concave towards the origin) is given in sufficient approximation by

$$B = \int_{-\infty}^{+\infty} \frac{\partial \gamma}{\partial x_1} dx_2,$$

while (73) and (70) give

$$\gamma = \sqrt{\left(-\frac{g_{44}}{g_{22}}\right)} = 1 - \frac{\alpha}{2r}\left(1 + \frac{x_2^2}{r^2}\right)$$

Carrying out the calculation, this gives

$$B = \frac{2\alpha}{\Delta} = \frac{\kappa M}{2\pi\Delta} \qquad (74)$$

According to this, a ray of light going past the sun undergoes a deflexion of 1·7"; and a ray going past the planet Jupiter a deflexion of about ·02".

If we calculate the gravitational field to a higher degree of approximation, and likewise with corresponding accuracy the orbital motion of a material point of relatively

* According to E. Freundlich, spectroscopical observations on fixed stars of certain types indicate the existence of an effect of this kind, but a crucial test of this consequence his not yet been made.

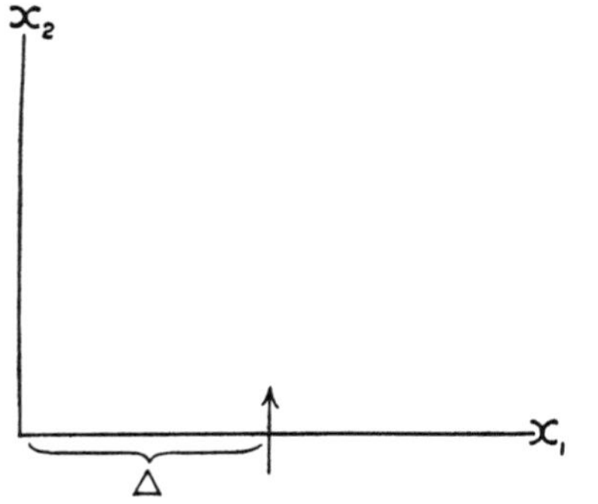

Fig. 8

infinitely small mass, we find a deviation of the following kind from the Kepler-Newton laws of planetary motion. The orbital ellipse of a planet undergoes a slow rotation, in the direction of motion, of amount

$$\varepsilon = 24\pi^3 \frac{a^2}{T^2 c^2 \left(1 - e^2\right)} \tag{75}$$

per revolution. In this formula *a* denotes the major semi-axis, *c* the velocity of light in the usual measurement, *e* the eccentricity, T the time of revolution in seconds.*

Calculation gives for the planet Mercury a rotation of the orbit of 43" per century, corresponding exactly to astronomical observation (Leverrier); for the astronomers have discovered in the motion of the perihelion of this planet, after allowing for disturbances by other planets, an inexplicable remainder of this magnitude.

HAMILTON'S PRINCIPLE AND THE GENERAL THEORY OF RELATIVITY

Translated from "Hamiltonsches Princip und allgemeine Relativitätstheorie," Sitzungsberichte der Preussischen Akad. Wissenschaften, 1916.

The general theory of relativity has recently been given in a particularly clear form by H. A. Lorentz and D. Hilbert,† who have deduced its equations from one single

* For the calculation I refer to the original papers: A. Einstein, Sitzungsber. d. Preuss. Akad. d. Wiss., 1915, p. 831; K. Schwarzschild, *ibid.*, 1916, p. 189.

† Four papers by Lorentz in the Publications of the Koninkl. Akad. van Wetensch. te Amsterdam, 1915 end 1916; D. Hilbert, Göttinger Nachr., 1915, Part 3.

principle of variation. The same thing will be done in the present paper. But my purpose here is to present the fundamental connexions in as perspicuous a manner as possible, and in as general terms as is permissible from the point of view of the general theory of relativity. In particular we shall make as few specializing assumptions as possible, in marked contrast to Hilbert's treatment of the subject. On the other hand, in antithesis to my own most recent treatment of the subject, there is to be complete liberty in the choice of the system of co-ordinates.

§ 1. THE PRINCIPLE OF VARIATION AND THE FIELD-EQUATIONS OF GRAVITATION AND MATTER

Let the gravitational field be described as usual by the tensor* of the $g_{\mu\nu}$ (or the $g^{\mu\nu}$); and matter, including the electromagnetic field, by any number of space-time functions $q_{(\rho)}$. How these functions may be characterized in the theory of invariants does not concern us. Further, let $\mathfrak{H}$ be a function of the

$$g^{\mu\nu}, g_{\sigma}^{\mu\nu}\left(=\frac{\partial g^{\mu\nu}}{\partial x_{\sigma}}\right) \text{ and } g_{\sigma\tau}^{\mu\nu}\left(=\frac{\partial^2 g^{\mu\nu}}{\partial x_{\sigma}\partial x_{\tau}}\right), \text{ the } q_{(\rho)} \text{ and } q_{(\rho)\alpha}\left(=\frac{\partial q_{(\rho)}}{\partial x_{\alpha}}\right).$$

The principle of variation

$$\delta\int \mathfrak{H}d\tau = 0 \tag{1}$$

then gives us as many differential equations as there are functions $g_{\mu\nu}$, and $q_{(\rho)}$ to be defined, if the $g^{\mu\nu}$ and $q_{(\rho)}$ are varied independently of one another, and in such a way that at the limits of integration the $\delta q_{(\rho)}$, $\delta g^{\mu\nu}$, and $\frac{\partial}{\partial x_{\sigma}}(\delta g_{\mu\nu})$ all vanish.

We will now assume that $\mathfrak{H}$ is linear in the $g_{\sigma\tau}$, and that the coefficients of the $g_{\sigma\tau}^{\mu\nu}$ depend only on the $g^{\mu\nu}$. We may then replace the principle of variation (1) by one which is more convenient for us. For by appropriate partial integration we obtain

$$\int \mathfrak{H}d\tau = \int \mathfrak{H}^{*}d\tau + F \tag{2}$$

where F denotes an integral over the boundary of the domain in question, and $\mathfrak{H}^{*}$ depends only on the $g^{\mu\nu}$, $g_{\sigma}^{\mu\nu}$, $q_{(\rho)}$, $q_{(\rho)\alpha}$, and no longer on the $g_{\sigma\tau}^{\mu\nu}$. From (2) we obtain, for such variations as are of interest to us,

$$\delta\int \mathfrak{H}d\tau = \delta\int \mathfrak{H}^{*}d\tau, \tag{3}$$

so that we may replace our principle of variation (1) by the more convenient form

$$\delta\int \mathfrak{H}^{*}d\tau = 0. \tag{1a}$$

* No use is made for the present of the tensor character of the $g_{\mu\nu}$.

By carrying out the variation of the $g^{\mu\nu}$ and the $q_{(\rho)}$, we obtain, as field-equations of gravitation and matter, the equations*

$$\frac{\partial}{\partial x_\alpha}\left(\frac{\partial \mathfrak{H}^*}{\partial g_\alpha^{\mu\nu}}\right) - \frac{\partial \mathfrak{H}^*}{\partial g^{\mu\nu}} = 0 \qquad (4)$$

$$\frac{\partial}{\partial x_\alpha}\left(\frac{\partial \mathfrak{H}^*}{\partial q_{(\rho)\alpha}}\right) - \frac{\partial \mathfrak{H}^*}{\partial q_{(\rho)}} = 0 \qquad (5)$$

§ 2. SEPARATE EXISTENCE OF THE GRAVITATIONAL FIELD

If we make no restrictive assumption as to the manner in which $\mathfrak{H}$ depends on the $g_{\mu\nu}$, $g_\sigma^{\mu\nu}$, $g_{\sigma\tau}^{\mu\nu}$, $q_{(\rho)}$, $q_{(\rho)\alpha}$, the energy-components cannot be divided into two parts, one belonging to the gravitational field, the other to matter. To ensure this feature of the theory, we make the following assumption

$$\mathfrak{H} = \mathfrak{G} + \mathfrak{M} \qquad (6)$$

where $\mathfrak{G}$ is to depend only on the $g^{\mu\nu}$, $g_\sigma^{\mu\nu}$, $g_{\sigma\tau}^{\mu\nu}$, and $\mathfrak{M}$ only on $g^{\mu\nu}$, $q_{(\rho)}$, $q_{(\rho)\alpha}$. Equations (4), (4a) then assume the form

$$\frac{\partial}{\partial x_\alpha}\left(\frac{\partial \mathfrak{G}^*}{\partial g_\alpha^{\mu\nu}}\right) - \frac{\partial \mathfrak{G}^*}{\partial g^{\mu\nu}} = \frac{\partial \mathfrak{M}}{\partial g^{\mu\nu}} \qquad (7)$$

$$\frac{\partial}{\partial x_\alpha}\left(\frac{\partial \mathfrak{M}}{\partial q_{(\rho)\alpha}}\right) - \frac{\partial \mathfrak{M}}{\partial q_{(\rho)}} = 0 \qquad (8)$$

Here $\mathfrak{G}^*$ stands in the same relation to $\mathfrak{G}$ as $\mathfrak{H}^*$ to $\mathfrak{H}$.

It is to be noted carefully that equations (8) or (5) would have to give way to others, if we were to assume $\mathfrak{M}$ or $\mathfrak{H}$ to be also dependent on derivatives of the $q_{(\rho)}$ of order higher than the first. Likewise it might be imaginable that the $q_{(\rho)}$ would have to be taken, not as independent of one another, but as connected by conditional equations. All this is of no importance for the following developments, as these are based solely on the equations (7), which have been found by varying our integral with respect to the $g^{\mu\nu}$.

§ 3. PROPERTIES OF THE FIELD EQUATIONS OF GRAVITATION CONDITIONED BY THE THEORY OF INVARIANTS

We now introduce the assumption that

$$ds^2 = g_{\mu\nu}dx_\mu dx_\nu \qquad (9)$$

* For brevity the summation symbols are omitted in the formulæ. Indices occurring twice in a term are always to be taken as summed. Thus in (4), for example, $\frac{\partial}{\partial x_\alpha}\left(\frac{\partial \mathfrak{H}^*}{\partial g_\alpha^{\mu\nu}}\right)$ denotes the term $\sum_\alpha \frac{\partial}{\partial x_\alpha}\left(\frac{\partial \mathfrak{H}^*}{\partial g_\alpha^{\mu\nu}}\right)$

is an invariant. This determines the transformational character of the $g_{\mu\nu}$. As to the transformational character of the $q_{(\rho)}$, which describe matter, we make no supposition. On the other hand, let the functions $H = \frac{\mathfrak{H}}{\sqrt{-g}}$, as well as $G = \frac{\mathfrak{G}}{\sqrt{-g}}$, and $M = \frac{\mathfrak{M}}{\sqrt{-g}}$, be invariants in relation to any substitutions and space-time co-ordinates. From these assumptions follows the general covariance of the equations (7) and (8), deduced from (1). It further follows that G (apart from a constant factor) must be equal to the scalar of Riemann's tensor of curvature; because there is no other invariant with the properties required for G.§ Thereby $\mathfrak{G}^*$ is also perfectly determined, and consequently the left-hand side of field equation (7) as well.†

From the general postulate of relativity there follow certain properties of the function $\mathfrak{G}^*$ which we shall now deduce. For this purpose we carry through an infinitesimal transformation of the co-ordinates, by setting

$$x'_\nu = x_\nu + \Delta x_\nu \tag{10}$$

where the Δx_ν are arbitrary, infinitely small functions of the co-ordinates, and x'_ν are the co-ordinates, in the new system, of the world-point having the co-ordinates x_ν in the original system. As for the co-ordinates, so too for any other magnitude ψ, a law of transformation holds good, of the type

$$\psi' = \psi + \Delta\psi,$$

where $\Delta\psi$ must always be expressible by the Δx_ν. From the covariant property of the $g^{\mu\nu}$ we easily deduce for the $g^{\mu\nu}$ and $g^{\mu\nu}_\sigma$ the laws of transformation

$$\Delta g^{\mu\nu} = g^{\mu\alpha}\frac{\partial(\Delta x_\nu)}{\partial x_\alpha} + g^{\nu\alpha}\frac{\partial(\Delta x_\mu)}{\partial x_\alpha} \tag{11}$$

$$\Delta g^{\mu\nu}_\sigma = \frac{\partial(\Delta g^{\mu\nu})}{\partial x_\sigma} - g^{\mu\nu}_\alpha\frac{\partial(\Delta x_\alpha)}{\partial x_\sigma} \tag{12}$$

Since $\mathfrak{G}^*$ depends only on the $g^{\mu\nu}$ and $g^{\mu\nu}_\sigma$, it is possible, with the help of (11) and (12), to calculate $\Delta\mathfrak{G}^*$. We thus obtain the equation

$$\sqrt{-g}\,\Delta\left(\frac{\mathfrak{G}^*}{\sqrt{-g}}\right) = \mathrm{S}^\nu_\sigma\frac{\partial(\Delta x_\sigma)}{\partial x_\nu} + 2\frac{\partial\mathfrak{G}^*}{\partial g^{\mu\sigma}_\alpha}g^{\mu\nu}\frac{\partial^2\Delta x_\sigma}{\partial x_\nu\partial x_\alpha}, \tag{13}$$

where for brevity we have set

$$\mathrm{S}^\nu_\sigma = 2\frac{\partial\mathfrak{G}^*}{\partial g^{\mu\sigma}}g^{\mu\nu} + 2\frac{\partial\mathfrak{G}^*}{\partial g^{\mu\sigma}_\alpha}g^{\mu\nu}_\alpha + \mathfrak{G}^*\delta^\nu_\sigma - \frac{\partial\mathfrak{G}^*}{\partial g^{\mu\alpha}_\nu}g^{\mu\alpha}_\sigma. \tag{14}$$

§ Herein is to be found the reason why the general postulate of relativity leads to a very definite theory of gravitation.
† By performing partial integration we obtain
$\mathfrak{G}^* = \sqrt{-g}\, g^{\mu\nu}[\{\mu\alpha, \beta\}\{\nu\beta, \alpha\} - \{\mu\nu, \alpha\}\{\alpha\beta, \beta\}]$.

From these two equations we draw two inferences which are important for what follows. We know that $\frac{\mathfrak{G}}{\sqrt{-g}}$ is an invariant with respect to any substitution, but we do not know this of $\frac{\mathfrak{G}^*}{\sqrt{-g}}$. It is easy to demonstrate, however, that the latter quantity is an invariant with respect to any *linear* substitutions of the co-ordinates. Hence it follows that the right side of (13) must always vanish if all $\frac{\partial^2 \Delta x_\sigma}{\partial x_\nu \partial x_\alpha}$ vanish. Consequently $\mathfrak{G}^*$ must satisfy the identity

$$S_\sigma^\nu \equiv 0 \tag{15}$$

If, further, we choose the Δx_ν so that they differ from zero only in the interior of a given domain, but in infinitesimal proximity to the boundary they vanish, then, with the transformation in question, the value of the boundary integral occurring in equation (2) does not change. Therefore $\Delta F = 0$, and, in consequence,*

$$\Delta \int \mathfrak{G} d\tau = \Delta \int \mathfrak{G}^* d\tau$$

But the left-hand side of the equation must vanish, since both $\frac{\mathfrak{G}}{\sqrt{-g}}$ and $\sqrt{-g}\,d\tau$ are invariants. Consequently the right-hand side also vanishes. Thus, taking (14), (15), and (16) into consideration, we obtain, in the first place, the equation

$$\int \frac{\partial \mathfrak{G}^*}{\partial g_\alpha^{\mu\sigma}} g^{\mu\nu} \frac{\partial^2(\Delta x_\sigma)}{\partial x_\nu \partial x_\alpha} d\tau = 0 \tag{16}$$

Transforming this equation by two partial integrations, and having regard to the liberty of choice of the Δx_σ, we obtain the identity

$$\frac{\partial^2}{\partial x_\nu \partial x_\alpha}\left(g^{\mu\nu} \frac{\partial \mathfrak{G}^*}{\partial g_\alpha^{\mu\sigma}}\right) \equiv 0 \tag{17}$$

From the two identities (16) and (17), which result from the invariance of $\frac{\mathfrak{G}}{\sqrt{-g}}$, and therefore from the postulate of general relativity, we now have to draw conclusions.

We first transform the field equations (7) of gravitation by mixed multiplication by $g^{\mu\sigma}$. We then obtain (by interchanging the indices σ and ν), as equivalents of the field equations (7) the equations

$$\frac{\partial}{\partial x_\alpha}\left(g^{\mu\nu} \frac{\partial \mathfrak{G}^*}{\partial g_\alpha^{\mu\sigma}}\right) = -(\mathfrak{T}_\sigma^\nu + \mathfrak{t}_\sigma^\nu) \tag{18}$$

where we have set

$$\mathfrak{T}_\sigma^\nu = -\frac{\partial \mathfrak{M}}{\partial g^{\mu\sigma}} g^{\mu\nu} \tag{19}$$

* By the introduction of the quantities $\mathfrak{G}$ and $\mathfrak{G}^*$ instead of $\mathfrak{H}$ and $\mathfrak{H}^*$.

$$t_\sigma^\nu = -\left(\frac{\partial \mathfrak{G}^*}{\partial g_\alpha^{\mu\sigma}} g_\alpha^{\mu\nu} + \frac{\partial \mathfrak{G}^*}{\partial g^{\mu\sigma}} g^{\mu\nu}\right) = \tfrac{1}{2}\left(\mathfrak{G}^* \delta_\sigma^\nu - \frac{\partial \mathfrak{G}^*}{\partial g_\nu^{\mu\alpha}} g_\sigma^{\mu\alpha}\right) \quad (20)$$

The last expression for t_μ^ν is vindicated by (14) and (15). By differentiation of (18) with respect to x_ν and summation for ν, there follows, in view of (17),

$$\frac{\partial}{\partial x_\nu}(\mathfrak{T}_\sigma^\nu + t_\sigma^\nu) = 0 \quad (21)$$

Equation (21) expresses the conservation of momentum and energy. We call $\mathfrak{T}_\sigma^\nu$ the components of the energy of matter, t_σ^ν the components of the energy of the gravitational field.

Having regard to (20), there follows from the field equations (7) of gravitation, by multiplication by $g_\sigma^{\mu\nu}$, and summation with respect to μ and ν,

$$\frac{\partial t_\sigma^\nu}{\partial x_\nu} + \tfrac{1}{2} g_\sigma^{\mu\nu} \frac{\partial \mathfrak{M}}{\partial g^{\mu\nu}} = 0,$$

or, in view of (19) and (21),

$$\frac{\partial \mathfrak{T}_\sigma^\nu}{\partial x_\nu} + \tfrac{1}{2} g_\sigma^{\mu\nu} \mathfrak{T}_{\mu\nu} = 0 \quad (22)$$

where $\mathfrak{T}_{\mu\nu}$ denotes the quantities $g_{\nu\sigma} \mathfrak{T}_\mu^\sigma$. These are four equations which the energy-components of matter have to satisfy.

It is to be emphasized that the (generally covariant) laws of conservation (21) and (22) are deduced from the field equations (7) of gravitation, in combination with the postulate of general covariance (relativity) *alone*, without using the field equations (8) for material phenomena.

COSMOLOGICAL CONSIDERATIONS ON THE GENERAL THEORY OF RELATIVITY

Translated from "Kosmologische Betrachtungen zur allgemeinen Relativitätstheorie," Sitzungsberichte der Preussischen Akad. d. Wissenschaften, 1917.

It is well known that Poisson's equation

$$\nabla^2 \phi = 4\pi K \rho \quad (1).$$

in combination with the equations of motion of a material point is not as yet a perfect substitute for Newton's theory of action at a distance. There is still to be taken into account the condition that at spatial infinity the potential ϕ tends toward a fixed limiting value. There is an analogous state of things in the theory of gravitation in general relativity. Here, too, we must supplement the differential equations by limiting conditions at spatial infinity, if we really have to regard the universe as being of infinite spatial extent.

In my treatment of the planetary problem I chose these limiting conditions in the form of the following assumption: it is possible to select a system of reference so that at spatial infinity all the gravitational potentials $g_{\mu\nu}$ become constant. But it is by no means evident *a priori* that we may lay down the same limiting conditions when we wish to take larger portions of the physical universe into consideration. In the following pages the reflexions will be given which, up to the present, I have made on this fundamentally important question.

§ 1. THE NEWTONIAN THEORY

It is well known that Newton's limiting condition of the constant limit for ϕ at spatial infinity leads to the view that the density of matter becomes zero at infinity. For we imagine that there may be a place in universal space round about which the gravitational field of matter, viewed on a large scale, possesses spherical symmetry. It then follows from Poisson's equation that, in order that ϕ may tend to a limit at infinity, the mean density ρ must decrease toward zero more rapidly than $1/r^2$ as the distance r from the centre increases.* In this sense, therefore, the universe according to Newton is finite, although it may possess an infinitely great total mass.

From this it follows in the first place that the radiation emitted by the heavenly bodies will, in part, leave the Newtonian system of the universe, passing radially outwards, to become ineffective and lost in the infinite. May not entire heavenly bodies fare likewise? It is hardly possible to give a negative answer to this question. For it follows from the assumption of a finite limit for ϕ at spatial infinity that a heavenly body with finite kinetic energy is able to reach spatial infinity by overcoming the Newtonian forces of attraction. By statistical mechanics this case must occur from time to time, as long as the total energy of the stellar system—transferred to one single star—is great enough to send that star on its journey to infinity, whence it never can return.

We might try to avoid this peculiar difficulty by assuming a very high value for the limiting potential at infinity. That would be a possible way, if the value of the gravitational potential were not itself necessarily conditioned by the heavenly bodies. The

* ρ is the mean density of matter, calculated for a region which is large as compared with the distance between neighbouring fixed stars, but small in comparison with the dimensions of the whole stellar system.

truth is that we are compelled to regard the occurrence of any great differences of potential of the gravitational field as contradicting the facts. These differences must really be of so low an order of magnitude that the stellar velocities generated by them do not exceed the velocities actually observed.

If we apply Boltzmann's law of distribution for gas molecules to the stars, by comparing the stellar system with a gas in thermal equilibrium, we find that the Newtonian stellar system cannot exist at all. For there is a finite ratio of densities corresponding to the finite difference of potential between the centre and spatial infinity. A vanishing of the density at infinity thus implies a vanishing of the density at the centre.

It seems hardly possible to surmount these difficulties on the basis of the Newtonian theory. We may ask ourselves the question whether they can be removed by a modification of the Newtonian theory. First of all we will indicate a method which does not in itself claim to be taken seriously; it merely serves as a foil for what is to follow. In place of Poisson's equation we write

$$\nabla^2\phi - \lambda\phi = 4\pi\kappa\rho \qquad (2)$$

where λ denotes a universal constant. If ρ_0 be the uniform density of distribution of mass, then

$$\phi = -\frac{4\pi\kappa}{\lambda}\rho_0 \qquad (3)$$

is a solution of equation (2). This solution would correspond to the case in which the matter of the fixed stars was distributed uniformly through space, if the density ρ_0 is equal to the actual mean density of the matter in the universe. The solution then corresponds to an infinite extension of the central space, filled uniformly with matter. If, without making any change in the mean density, we imagine matter to be non-uniformly distributed locally, there will be, over and above the ϕ with the constant value of equation (3), an additional ϕ, which in the neighbourhood of denser masses will so much the more resemble the Newtonian field as $\lambda\phi$ is smaller in comparison with $4\pi\kappa\rho$.

A universe so constituted would have, with respect to its gravitational field, no centre. A decrease of density in spatial infinity would not have to be assumed, but both the mean potential and mean density would remain constant to infinity. The conflict with statistical mechanics which we found in the case of the Newtonian theory is not repeated. With a definite but extremely small density, matter is in equilibrium, without any internal material form (pressures) being required to maintain equilibrium.

§ 2. THE BOUNDARY CONDITIONS ACCORDING TO THE GENERAL THEORY OF RELATIVITY

In the present paragraph I shall conduct the reader over the road that I have myself travelled, rather a rough and winding road, because otherwise I cannot hope that he will take much interest in the result at the end of the journey. The conclusion I shall arrive at is that the field equations of gravitation which I have championed hitherto still need a slight modification, so that on the basis of the general theory of relativity those fundamental difficulties may be avoided which have been set forth in § 1 as confronting the Newtonian theory. This modification corresponds perfectly to the transition from Poisson's equation (1) to equation (2) of § 1. We finally infer that boundary conditions in spatial infinity fall away altogether, because the universal continuum in respect of its spatial dimensions is to be viewed as a self-contained continuum of finite spatial (three-dimensional) volume.

The opinion which I entertained until recently, as to the limiting conditions to be laid down in spatial infinity, took its stand on the following considerations. In a consistent theory of relativity there can be no inertia *relatively to "space,"* but only an inertia of masses *relatively to one another.* If, therefore, I have a mass at a sufficient distance from all other masses in the universe, its inertia must fall to zero. We will try to formulate this condition mathematically.

According to the general theory of relativity the negative momentum is given by the first three components, the energy by the last component of the covariant tensor multiplied by $\sqrt{-g}$

$$m\sqrt{-g}\quad g_{\mu\alpha}\frac{dx_\alpha}{ds} \tag{4}$$

where, as always, we set

$$ds^2 = g_{\mu\nu}dx_\mu dx_\nu \tag{5}$$

In the particularly perspicuous case of the possibility of choosing the system of co-ordinates so that the gravitational field at every point is spatially isotropic, we have more simply

$$ds^2 = -\mathrm{A}\left(dx_1^2 + dx_2^2 + dx_3^2\right) + \mathrm{B}dx_4^2$$

If, moreover, at the same time

$$\sqrt{-g} = 1 = \sqrt{\mathrm{A}^3\mathrm{B}}$$

we obtain from (4), to a first approximation for small velocities,

$$m\frac{A}{\sqrt{B}}\frac{dx_1}{dx_4}, m\frac{A}{\sqrt{B}}\frac{dx_2}{dx_4}, m\frac{A}{\sqrt{B}}\frac{dx_3}{dx_4}$$

for the components of momentum, and for the energy (in the static case)

$$m\sqrt{B}\ .$$

From the expressions for the momentum, it follows that $m\frac{A}{\sqrt{B}}$ plays the part of the rest mass. As m is a constant peculiar to the point of mass, independently of its position, this expression, if we retain the condition $\sqrt{-g} = 1$ at spatial infinity, can vanish only when A diminishes to zero, while B increases to infinity. It seems, therefore, that such a degeneration of the co-efficients $g_{\mu\nu}$ is required by the postulate of relativity of all inertia. This requirement implies that the potential energy $m\sqrt{B}$ becomes infinitely great at infinity. Thus a point of mass can never leave the system; and a more detailed investigation shows that the same thing applies to light-rays. A system of the universe with such behaviour of the gravitational potentials at infinity would not therefore run the risk of wasting away which was mooted just now in connexion with the Newtonian theory.

I wish to point out that the simplifying assumptions as to the gravitational potentials on which this reasoning is based, have been introduced merely for the sake of lucidity. It is possible to find general formulations for the behaviour of the $g_{\mu\nu}$ at infinity which express the essentials of the question without further restrictive assumptions.

At this stage, with the kind assistance of the mathematician J. Grommer, I investigated centrally symmetrical, static gravitational fields, degenerating at infinity in the way mentioned. The gravitational potentials $g_{\mu\nu}$ were applied, and from them the energy-tensor $T_{\mu\nu}$ of matter was calculated on the basis of the field equations of gravitation. But here it proved that for the system of the fixed stars no boundary conditions of the kind can come into question at all, as was also rightly emphasized by the astronomer de Sitter recently.

For the contravariant energy-tensor $T^{\mu\nu}$ of ponderable matter is given by

$$T^{\mu\nu} = \rho\frac{dx_\mu}{ds}\frac{dx_\nu}{ds},$$

where ρ is the density of matter in natural measure. With an appropriate choice of the system of co-ordinates the stellar velocities are very small in comparison with that of light. We may, therefore, substitute $\sqrt{g_{44}}\,dx_4$ for ds. This shows us that all components of $T^{\mu\nu}$ must be very small in comparison with the last component T^{44}. But it was quite impossible to reconcile this condition with the chosen boundary conditions. In the retrospect this result does not appear astonishing. The fact of the small velocities of the stars allows the conclusion that wherever there are fixed stars, the gravitational potential (in our case $\sqrt{B}$) can never be much greater than here on earth. This follows from statistical reasoning, exactly as in the case of the Newtonian theory. At any rate, our calculations have convinced me that such conditions of degeneration for the $g_{\mu\nu}$ in spatial infinity may not be postulated.

After the failure of this attempt, two possibilities next present themselves.

(*a*) We may require, as in the problem of the planets, that, with a suitable choice of the system of reference, the $g_{\mu\nu}$ in spatial infinity approximate to the values

$$\begin{matrix} -1 & 0 & 0 & 0 \\ 0 & -1 & 0 & 0 \\ 0 & 0 & -1 & 0 \\ 0 & 0 & 0 & 1 \end{matrix}$$

(*b*) We may refrain entirely from laying down boundary conditions for spatial infinity claiming general validity; but at the spatial limit of the domain under consideration we have to give the $g_{\mu\nu}$ separately in each individual case, as hitherto we were accustomed to give the initial conditions for time separately.

The possibility (*b*) holds out no hope of solving the problem, but amounts to giving it up. This is an incontestable position, which is taken up at the present time by de Sitter.* But I must confess that such a complete resignation in this fundamental question is for me a difficult thing. I should not make up my mind to it until every effort to make headway toward a satisfactory view had proved to be vain.

Possibility (*a*) is unsatisfactory in more respects than one. In the first place those boundary conditions pre-suppose a definite choice of the system of reference, which is contrary to the spirit of the relativity principle. Secondly, if we adopt this view, we fail to comply with the requirement of the relativity of inertia. For the inertia of a material point of mass m (in natural measure) depends upon the $g_{\mu\nu}$; but these differ but little from their postulated values, as given above, for spatial infinity. Thus inertia would indeed be *influenced*, but would not be *conditioned* by matter (present in finite space). If only one single point of mass were present, according to this view, it would possess inertia, and in fact an inertia almost as great as when it is surrounded by the other masses of the actual universe. Finally, those statistical objections must be raised against this view which were mentioned in respect of the Newtonian theory.

From what has now been said it will be seen that I have not succeeded in formulating boundary conditions for spatial infinity. Nevertheless, there is still a possible way out, without resigning as suggested under (*b*). For if it were possible to regard the universe as a continuum which is *finite (closed) with respect to its spatial dimensions*, we should have no need at all of any such boundary conditions. We shall proceed to show that both the general postulate of relativity and the fact of the small stellar velocities

* de Sitter, Akad. van Wetensch. te Amsterdam, 8 Nov., 1916.

are compatible with the hypothesis of a spatially finite universe; though certainly, in order to carry through this idea, we need a generalizing modification of the field equations of gravitation.

§ 3. THE SPATIALLY FINITE UNIVERSE WITH A UNIFORM DISTRIBUTION OF MATTER

According to the general theory of relativity the metrical character (curvature) of the four-dimensional space-time continuum is defined at every point by the matter at that point and the state of that matter. Therefore, on account of the lack of uniformity in the distribution of matter, the metrical structure of this continuum must necessarily be extremely complicated. But if we are concerned with the structure only on a large scale, we may represent matter to ourselves as being uniformly distributed over enormous spaces, so that its density of distribution is a variable function which varies extremely slowly. Thus our procedure will somewhat resemble that of the geodesists who, by means of an ellipsoid, approximate to the shape of the earth's surface, which on a small scale is extremely complicated.

The most important fact that we draw from experience as to the distribution of matter is that the relative velocities of the stars are very small as compared with the velocity of light. So I think that for the present we may base our reasoning upon the following approximative assumption. There is a system of reference relatively to which matter may be looked upon as being permanently at rest. With respect to this system, therefore, the contravariant energy-tensor $T^{\mu\nu}$ of matter is, by reason of (5), of the simple form

$$\left.\begin{matrix} 0 & 0 & 0 & 0 \\ 0 & 0 & 0 & 0 \\ 0 & 0 & 0 & 0 \\ 0 & 0 & 0 & \rho \end{matrix}\right\} \qquad (6)$$

The scalar ρ of the (mean) density of distribution may be *a priori* a function of the space co-ordinates. But if we assume the universe to be spatially finite, we are prompted to the hypothesis that ρ is to be independent of locality. On this hypothesis we base the following considerations.

As concerns the gravitational field, it follows from the equation of motion of the material point

$$\frac{d^2 x_\nu}{ds^2} + \{\alpha\beta,\ \nu\}\frac{dx_\alpha}{ds}\frac{dx_\beta}{ds} = 0$$

that a material point in a static gravitational field can remain at rest only when g_{44} is independent of locality. Since, further, we presuppose independence of the time co-ordinate x_4 for all magnitudes, we may demand for the required solution that, for all x_ν,

$$g_{44} = 1 \tag{7}$$

Further, as always with static problems, we shall have to set

$$g_{14} = g_{24} = g_{34} = 0 \tag{8}$$

It remains now to determine those components of the gravitational potential which define the purely spatial-geometrical relations of our continuum ($g_{11}, g_{12}, \ldots g_{33}$). From our assumption as to the uniformity of distribution of the masses generating the field, it follows that the curvature of the required space must be constant. With this distribution of mass, therefore, the required finite continuum of the x_1, x_2, x_3, with constant x_4, will be a spherical space.

We arrive at such a space, for example, in the following way. We start from a Euclidean space of four dimensions, $\xi_1, \xi_2, \xi_3, \xi_4$, with a linear element $d\sigma$; let, therefore,

$$d\sigma^2 = d\xi_1^2 + d\xi_2^2 + d\xi_3^2 + d\xi_4^2 \tag{9}$$

In this space we consider the hyper-surface

$$R^2 = \xi_1^2 + \xi_2^2 + \xi_3^2 + \xi_4^2, \tag{10}$$

where R denotes a constant. The points of this hyper-surface form a three-dimensional continuum, a spherical space of radius of curvature R.

The four-dimensional Euclidean space with which we started serves only for a convenient definition of our hyper-surface. Only those points of the hyper-surface are of interest to us which have metrical properties in agreement with those of physical space with a uniform distribution of matter. For the description of this three-dimensional continuum we may employ the co-ordinates ξ_1, ξ_2, ξ_3 (the projection upon the hyper-plane $\xi_4 = 0$) since, by reason of (10), ξ_4 can be expressed in terms of ξ_1, ξ_2, ξ_3. Eliminating ξ_4 from (9), we obtain for the linear element of the spherical space the expression

$$\left.\begin{aligned} d\sigma^2 &= \gamma_{\mu\nu} d\xi_\mu d\xi_\nu \\ \gamma_{\mu\nu} &= \delta_{\mu\nu} + \frac{\xi_\mu \xi_\nu}{R^2 - \rho^2} \end{aligned}\right\} \tag{11}$$

where $\delta_{\mu\nu} = 1$, if $\mu = \nu$; $\delta_{\mu\nu} = 0$, if $\mu \neq \nu$, and $\rho^2 = \xi_1^2 + \xi_2^2 + \xi_3^2$. The co-ordinates chosen are convenient when it is a question of examining the environment of one of the two points $\xi_1 = \xi_2 = \xi_3 = 0$.

Now the linear element of the required four-dimensional space-time universe is also given us. For the potential $g_{\mu\nu}$, both indices of which differ from 4, we have to set

$$g_{\mu\nu} = -\left(\delta_{\mu\nu} + \frac{x_\mu x_\nu}{R^2 - (x_1^2 + x_2^2 + x_3^2)}\right) \tag{12}$$

which equation, in combination with (7) and (8), perfectly defines the behaviour of measuring-rods, clocks, and light-rays.

§ 4. ON AN ADDITIONAL TERM FOR THE FIELD EQUATIONS OF GRAVITATION

My proposed field equations of gravitation for any chosen system of co-ordinates run as follows:—

$$\left.\begin{aligned} G_{\mu\nu} &= -\kappa\left(T_{\mu\nu} - \frac{1}{2}g_{\mu\nu}T\right), \\ G_{\mu\nu} &= -\frac{\partial}{\partial x_\alpha}\{\mu\nu, \alpha\} + \{\mu\nu, \beta\}\{\nu\beta, \alpha\} \\ &\quad + \frac{\partial^2 \log\sqrt{-g}}{\partial x_\mu \partial x_\nu} - \{\mu\nu, \alpha\}\frac{\partial \log\sqrt{-g}}{\partial x_\alpha} \end{aligned}\right\} \qquad (13)$$

The system of equations (13) is by no means satisfied when we insert for the $g_{\mu\nu}$ the values given in (7), (8), and (12), and for the (contravariant) energy-tensor of matter the values indicated in (6). It will be shown in the next paragraph how this calculation may conveniently be made. So that, if it were certain that the field equations (13) which I have hitherto employed were the only ones compatible with the postulate of general relativity, we should probably have to conclude that the theory of relativity does not admit the hypothesis of a spatially finite universe.

However, the system of equations (14) allows a readily suggested extension which is compatible with the relativity postulate, and is perfectly analogous to the extension of Poisson's equation given by equation (2). For on the left-hand side of field equation (13) we may add the fundamental tensor $g_{\mu\nu}$, multiplied by a universal constant, $-\lambda$, at present unknown, without destroying the general covariance. In place of field equation (13) we write

$$G_{\mu\nu} - \lambda g_{\mu\nu} = -\kappa\left(T_{\mu\nu} - \frac{1}{2}g_{\mu\nu}T\right) \qquad (13a)$$

This field equation, with λ sufficiently small, is in any case also compatible with the facts of experience derived from the solar system. It also satisfies laws of conservation of momentum and energy, because we arrive at (13a) in place of (13) by introducing into Hamilton's principle, instead of the scalar of Riemann's tensor, this scalar increased by a universal constant; and Hamilton's principle, of course, guarantees the validity of laws of conservation. It will be shown in § 5 that field equation (13a) is compatible with our conjectures on field and matter.

§ 5. CALCULATION AND RESULT

Since all points of our continuum are on an equal footing, it is sufficient to carry through the calculation for *one* point, e.g. for one of the two points with the co-ordinates

$$x_1 = x_2 = x_3 = x_4 = 0.$$

Then for the $g_{\mu\nu}$ in (13a) we have to insert the values

$$\begin{matrix} -1 & 0 & 0 & 0 \\ 0 & -1 & 0 & 0 \\ 0 & 0 & -1 & 0 \\ 0 & 0 & 0 & 1 \end{matrix}$$

wherever they appear differentiated only once or not at all. We thus obtain in the first place

$$G_{\mu\nu} = \frac{\partial}{\partial x_1}[\mu\nu, 1] + \frac{\partial}{\partial x_2}[\mu\nu, 2] + \frac{\partial}{\partial x_3}[\mu\nu, 3] + \frac{\partial^2 \log \sqrt{-g}}{\partial x_\mu \partial x_\nu}.$$

From this we readily discover, taking (7), (8), and (13) into account, that all equations (13a) are satisfied if the two relations

$$-\frac{2}{R^2} + \lambda = -\frac{\kappa\rho}{2}, \quad -\lambda = -\frac{\kappa\rho}{2},$$

or

$$\lambda = \frac{\kappa\rho}{2} = \frac{1}{R^2} \tag{14}$$

are fulfilled.

Thus the newly introduced universal constant λ defines both the mean density of distribution ρ which can remain in equilibrium and also the radius R and the volume $2\pi^2 R^3$ of spherical space. The total mass M of the universe, according to our view, is finite, and is in fact

$$M = \rho \cdot 2\pi^2 R^3 = 4\pi^2 \frac{R}{\kappa} = \pi^2 \sqrt{\frac{32}{\kappa^3 \rho}} \tag{15}$$

Thus the theoretical view of the actual universe, if it is in correspondence with our reasoning, is the following. The curvature of space is variable in time and place, according to the distribution of matter, but we may roughly approximate to it by means of a spherical space. At any rate, this view is logically consistent, and from the standpoint of the general theory of relativity lies nearest at hand; whether, from the standpoint of present astronomical knowledge, it is tenable, will not here be discussed. In order to arrive at this consistent view, we admittedly had to introduce an extension of the field equations of gravitation which is not justified by our actual knowledge of gravitation.

It is to be emphasized, however, that a positive curvature of space is given by our results, even if the supplementary term is not introduced. That term is necessary only for the purpose of making possible a quasi-static distribution of matter, as required by the fact of the small velocities of the stars.

DO GRAVITATIONAL FIELDS PLAY AN ESSENTIAL PART IN THE STRUCTURE OF THE ELEMENTARY PARTICLES OF MATTER?

Translated from "Spielen Gravitationsfelder im Aufber der materiellen Elementarteilchen eine wesentliche Rolle?" Sitzungsberichte der Preussischen Akad. d. Wissenschaften,
1919.

Neither the Newtonian nor the relativistic theory of gravitation has so far led to any advance in the theory of the constitution of matter. In view of this fact it will be shown in the following pages that there are reasons for thinking that the elementary formations which go to make up the atom are held together by gravitational forces.

§ 1. DEFECTS OF THE PRESENT VIEW

Great pains have been taken to elaborate a theory which will account for the equilibrium of the electricity constituting the electron. G. Mie, in particular, has devoted deep researches to this question. His theory, which has found considerable support among theoretical physicists, is based mainly on the introduction into the energy-tensor of supplementary terms depending on the components of the electro-dynamic potential, in addition to the energy terms of the Maxwell-Lorentz theory. These new terms, which in outside space are unimportant, are nevertheless effective in the interior of the electrons in maintaining equilibrium against the electric form of repulsion. In spite of the beauty of the formal structure of this theory, as erected by Mie, Hilbert, and Weyl, its physical results have hitherto been unsatisfactory. On the one hand the multiplicity of possibilities is discouraging, and on the other hand those additional terms have not as yet allowed themselves to be framed in such a simple form that the solution could be satisfactory.

So far the general theory of relativity has made no change in this state of the question. If we for the moment disregard the additional cosmological term, the field equations take the form

$$G_{\mu\nu} - \frac{1}{2} g_{\mu\nu} G = -\kappa T_{\mu\nu} \tag{1}$$

where $G_{\mu\nu}$ denotes the contracted Riemann tensor of curvature, G the scalar of curvature formed by repeated contraction, and $T_{\mu\nu}$ the energy-tensor of "matter." The assumption that the $T_{\mu\nu}$ do *not* depend on the derivatives of the $g_{\mu\nu}$ is in keeping with the historical development of these equations. For these quantities are, of course, the energy-components in the sense of the special theory of relativity, in which variable $g_{\mu\nu}$ do not occur. The second term on the left-hand side of the equation is so chosen that the divergence of the left-hand side of (1) vanishes identically, so that taking the divergence of (1), we obtain the equation

$$\frac{\partial \mathfrak{T}_{\mu}^{\sigma}}{\partial x_{\sigma}} + \tfrac{1}{2} g_{\mu}^{\sigma\tau} \mathfrak{T}_{\sigma\tau} = 0 \tag{2}$$

which in the limiting case of the special theory of relativity gives the complete equations of conservation

$$\frac{\partial T_{\mu\nu}}{\partial x_{\nu}} = 0.$$

Therein lies the physical foundation for the second term of the left-hand side of (1). It is by no means settled *a priori* that a limiting transition of this kind has any possible meaning. For if gravitational fields do play an essential part in the structure of the particles of matter, the transition to the limiting case of constant $g_{\mu\nu}$ would, for them, lose its justification, for indeed, with constant $g_{\mu\nu}$ there could not be any particles of matter. So if we wish to contemplate the possibility that gravitation may take part in the structure of the fields which constitute the corpuscles, we cannot regard equation (1) as confirmed.

Placing in (1) the Maxwell-Lorentz energy-components of the electromagnetic field $\phi_{\mu\nu}$,

$$T_{\mu\nu} = \frac{1}{4} g_{\mu\nu} \phi_{\sigma\tau} \phi^{\sigma\tau} - \phi_{\mu\sigma} \phi_{\nu\tau} g^{\sigma\tau} \tag{3}$$

we obtain for (2), by taking the divergence, and after some reduction,*

$$\phi_{\mu\sigma} \mathfrak{J}^{\sigma} = 0 \tag{4}$$

where, for brevity, we have set

$$\frac{\partial}{\partial x_{\tau}} \left(\sqrt{-g} \phi_{\mu\nu} g^{\mu\sigma} g^{\nu\tau} \right) = \frac{\partial \mathfrak{f}^{\sigma\tau}}{\partial x_{\tau}} = \mathfrak{J}^{\sigma} \tag{5}$$

* Cf. e.g. A. Einstein, Sitzungsber. d. Preuss. Akad. d. Wiss., 1916, pp. 187, 188.

In the calculation we have employed the second of Maxwell's systems of equations

$$\frac{\partial \phi_{\mu\nu}}{\partial x_\rho} + \frac{\partial \phi_{\nu\rho}}{\partial x_\mu} + \frac{\partial \phi_{\rho\mu}}{\partial x_\nu} = 0 \qquad (6)$$

We see from (4) that the current-density $\mathfrak{J}^\sigma$ must everywhere vanish. Therefore, by equation (1), we cannot arrive at a theory of the electron by restricting ourselves to the electromagnetic components of the Maxwell-Lorentz theory, as has long been known. Thus if we hold to (1) we are driven on to the path of Mie's theory.*

Not only the problem of matter, but the cosmological problem as well, leads to doubt as to equation (1). As I have shown in the previous paper, the general theory of relativity requires that the universe be spatially finite. But this view of the universe necessitated an extension of equations (1), with the introduction of a new universal constant λ, standing in a fixed relation to the total mass of the universe (or, respectively, to the equilibrium density of matter). This is gravely detrimental to the formal beauty of the theory.

§ 2. THE FIELD EQUATIONS FREED OF SCALARS

The difficulties set forth above are removed by setting in place of field equations (1) the field equations

$$G_{\mu\nu} - \frac{1}{4} g_{\mu\nu} G = -\kappa T_{\mu\nu} \qquad (1a)$$

where $T_{\mu\nu}$ denotes the energy-tensor of the electromagnetic field given by (3).

The formal justification for the factor $-{}^1/_4$ in the second term of this equation lies in its causing the scalar of the left-hand side,

$$g^{\mu\nu}\left(G_{\mu\nu} - \frac{1}{4} g_{\mu\nu} G\right),$$

to vanish identically, as the scalar $g^{\nu\mu}T_{\mu\nu}$ of the right-hand side does by reason of (3). If we had reasoned on the basis of equations (1) instead of (1a), we should, on the contrary, have obtained the condition $G = 0$, which would have to hold good everywhere for the $g_{\mu\nu}$, independently of the electric field. It is clear that the system of equations [(1a), (3)] is a consequence of the system [(1), (3)], but not conversely.

We might at first sight feel doubtful whether (1a) together with (6) sufficiently define the entire field. In a generally relativistic theory we need $n - 4$ differential equations, independent of one another, for the definition of n independent variables, since in the solution, on account of the liberty of choice of the co-ordinates, four quite arbitrary functions of all co-ordinates must naturally occur. Thus to define the sixteen independent quantities $g_{\mu\nu}$ and $\phi_{\mu\nu}$ we require twelve equations, all independent of one another. But as it happens, nine of the equations (1a), and three of the equations (6) are independent of one another.

* Cf. D. Hilbert, Göttinger Nachr., 20 Nov., 1915.

Forming the divergence of (1a), and taking into account that the divergence of $G_{\mu\nu} - \frac{1}{2} g_{\mu\nu} G$ vanishes, we obtain

$$\phi_{\sigma\alpha} J^{\alpha} + \frac{1}{4\kappa} \frac{\partial G}{\partial x_{\sigma}} = 0 \qquad (4a)$$

From this we recognize first of all that the scalar of curvature G in the four-dimensional domains in which the density of electricity vanishes, is constant. If we assume that all these parts of space are connected, and therefore that the density of electricity differs from zero only in separate "world-threads," then the scalar of curvature, everywhere outside these world-threads, possesses a constant value G_0. But equation (4a) also allows an important conclusion as to the behaviour of G within the domains having a density of electricity other than zero. If, as is customary, we regard electricity as a moving density of charge, by setting

$$J^{\sigma} = \frac{\mathfrak{J}^{\sigma}}{\sqrt{-g}} = \rho \frac{dx_{\sigma}}{ds}, \qquad (7)$$

we obtain from (4a) by inner multiplication by J^{σ}, on account of the antisymmetry of $\phi_{\mu\nu}$, the relation

$$\frac{\partial G}{\partial x_{\sigma}} \frac{dx_{\sigma}}{ds} = 0 \qquad (8)$$

Thus the scalar of curvature is constant on every world-line of the motion of electricity. Equation (4a) can be interpreted in a graphic manner by the statement: The scalar of curvature plays the part of a negative pressure which, outside of the electric corpuscles, has a constant value G_0. In the interior of every corpuscle there subsists a negative pressure (positive $G - G_0$) the fall of which maintains the electrodynamic force in equilibrium. The minimum of pressure, or, respectively, the maximum of the scalar of curvature, does not change with time in the interior of the corpuscle.

We now write the field equations (1a) in the form

$$\left(G_{\mu\nu} - \frac{1}{2} g_{\mu\nu} G\right) + \frac{1}{4} g_{\mu\nu} G_0 = -\kappa\left(T_{\mu\nu} + \frac{1}{4\kappa} g_{\mu\nu}(G - G_0)\right) \qquad (9)$$

On the other hand, we transform the equations supplied with the cosmological term as already given

$$G_{\mu\nu} - \lambda g_{\mu\nu} = -\kappa\left(T_{\mu\nu} - \frac{1}{2} g_{\mu\nu} T\right)$$

Subtracting the scalar equation multiplied by $^1/_2$, we next obtain

$$\left(G_{\mu\nu} - \frac{1}{2} g_{\mu\nu} G\right) + g_{\mu\nu}\lambda = -\kappa T_{\mu\nu}.$$

Now in regions where only electrical and gravitational fields are present, the right-hand side of this equation vanishes. For such regions we obtain, by forming the scalar,

$$-G + 4\lambda = 0.$$

In such regions, therefore, the scalar of curvature is constant, so that λ may be replaced by $^1/_4G_0$. Thus we may write the earlier field equation (1) in the form

$$G_{\mu\nu} - \frac{1}{2}g_{\mu\nu}G + \frac{1}{4}g_{\mu\nu}G_0 = -\kappa T_{\mu\nu} \tag{10}$$

Comparing (9) with (10), we see that there is no difference between the new field equations and the earlier ones, except that instead of $T_{\mu\nu}$ as tensor of "gravitating mass" there now occurs $T_{\mu\nu} + \frac{1}{4\kappa}g_{\mu\nu}(G - G_0)$ which is independent of the scalar of curvature. But the new formulation has this great advantage, that the quantity λ appears in the fundamental equations as a constant of integration, and no longer as a universal constant peculiar to the fundamental law.

§ 3. ON THE COSMOLOGICAL QUESTION

The last result already permits the surmise that with our new formulation the universe may be regarded as spatially finite, without any necessity for an additional hypothesis. As in the preceding paper I shall again show that with a uniform distribution of matter, a spherical world is compatible with the equations.

In the first place we set

$$ds^2 = -\gamma_{ik}dx_i dx_k + dx_4^2 \quad (i, k = 1, 2, 3) \tag{11}$$

Then if P_{ik} and P are, respectively, the curvature tensor of the second rank and the curvature scalar in three-dimensional space, we have

$$\begin{aligned} G_{ik} &= P_{ik} \quad (i, k = 1, 2, 3) \\ G_4^i &= G_{4i} = G_{44} = 0 \\ G &= -P \\ -g &= \gamma. \end{aligned}$$

It therefore follows for our case that

$$\begin{aligned} G_{ik} - \frac{1}{2}g_{ik}G &= P_{ik} - \frac{1}{2}\gamma_{ik}P \quad (i, k = 1, 2, 3) \\ G_{44} - \frac{1}{2}g_{44}G &= \frac{1}{2}P. \end{aligned}$$

We pursue our reflexions, from this point on, in two ways. Firstly, with the support of equation (1a). Here $T_{\mu\nu}$ denotes the energy-tensor of the electro-magnetic field, arising from the electrical particles constituting matter. For this field we have everywhere

$$\mathfrak{T}_1^1 + \mathfrak{T}_2^2 + \mathfrak{T}_3^3 + \mathfrak{T}_4^4 = 0.$$

The individual $\mathfrak{T}_\mu^\nu$ are quantities which vary rapidly with position; but for our purpose we no doubt may replace them by their mean values. We therefore have to choose

$$\left.\begin{aligned}\mathfrak{T}_1^1 = \mathfrak{T}_2^2 = \mathfrak{T}_3^3 = -\tfrac{1}{3}\mathfrak{T}_4^4 = \text{const.}\\ \mathfrak{T}_\mu^\nu = 0 \ (\text{for } \mu \neq \nu),\end{aligned}\right\} \quad (12)$$

and therefore

$$\mathrm{T}_{ik} = \tfrac{1}{3}\frac{\mathfrak{T}_4^4}{\sqrt{\gamma}}\gamma_{ik}, \ \mathrm{T}_{44} = \frac{\mathfrak{T}_4^4}{\sqrt{\gamma}}.$$

In consideration of what has been shown hitherto, we obtain in place of (1a)

$$\mathrm{P}_{ik} - \tfrac{1}{4}\gamma_{ik}\mathrm{P} = -\tfrac{1}{3}\gamma_{ik}\frac{\kappa\mathfrak{T}_4^4}{\sqrt{\gamma}} \quad (13)$$

$$\tfrac{1}{4}\mathrm{P} = -\frac{\kappa\mathfrak{T}_4^4}{\sqrt{\gamma}} \quad (14)$$

The scalar of equation (13) agrees with (14). It is on this account that our fundamental equations permit the idea of a spherical universe. For from (13) and (14) follows

$$\mathrm{P}_{ik} + \frac{4}{3}\frac{\kappa\mathfrak{T}_4^4}{\sqrt{\gamma}}\gamma_{ik} = 0 \quad (15)$$

and it is known* that this system is satisfied by a (three-dimensional) spherical universe.

But we may also base our reflexions on the equations (9). On the right-hand side of (9) stand those terms which, from the phenomenological point of view, are to be replaced by the energy-tensor of matter; that is, they are to be replaced by

$$\begin{matrix} 0 & 0 & 0 & 0 \\ 0 & 0 & 0 & 0 \\ 0 & 0 & 0 & 0 \\ 0 & 0 & 0 & \rho \end{matrix}$$

where ρ denotes the mean density of matter assumed to be at rest. We thus obtain the equations

$$\mathrm{P}_{ik} - \frac{1}{2}\gamma_{ik}\mathrm{P} - \frac{1}{4}\gamma_{ik}\mathrm{G}_0 = 0 \quad (16)$$

$$\frac{1}{2}\mathrm{P} + \frac{1}{4}\mathrm{G}_0 = -\kappa\rho \quad (17)$$

From the scalar of equation (16) and from (17) we obtain

$$\mathrm{G}_0 = -\frac{2}{3}\mathrm{P} = 2\kappa\rho \quad (18)$$

and consequently from (16)

$$\mathrm{P}_{ik} - \kappa\rho\gamma_{ik} = 0 \quad (19)$$

which equation, with the exception of the expression for the co-efficient, agrees with

* Cf. H. Weyl, "Raum, Zeit, Materie," § 33.

(15). By comparison we obtain

$$\mathfrak{T}_4^4 = \tfrac{3}{4}\rho\sqrt{\gamma} \tag{20}$$

This equation signifies that of the energy constituting matter three-quarters is to be ascribed to the electromagnetic field, and one-quarter to the gravitational field.

§ 4. CONCLUDING REMARKS

The above reflexions show the possibility of a theoretical construction of matter out of gravitational field and electromagnetic field alone, without the introduction of hypothetical supplementary terms on the lines of Mie's theory. This possibility appears particularly promising in that it frees us from the necessity of introducing a special constant λ for the solution of the cosmological problem. On the other hand, there is a peculiar difficulty. For, if we specialize (1) for the spherically symmetrical static case we obtain one equation too few for defining the $g_{\mu\nu}$ and $\phi_{\mu\nu}$, with the result that any *spherically symmetrical distribution* of electricity appears capable of remaining in equilibrium. Thus the problem of the constitution of the elementary quanta cannot yet be solved on the immediate basis of the given field equations.

Acknowledgments

This book would not have been possible without the help of a number of talented people who made different contributions at various stages of the book's development. Among those deserving special thanks are Michael Rosin, a consultant to Running Press, Gil King, and Mrs. Karen Sime, assistant to Professor Stephen Hawking.

Thanks are also due to Carlo DeVito, Kathleen Greczylo, Kelly Pennick, Bill Jones, Deborah Grandinetti, Sarah O'Brien, and Jesse Kinder.

About Stephen Hawking

Stephen Hawking is considered the most brilliant theoretical physicist since Einstein. He has also done much to popularize science. His book, *A Brief History of Time*, sold more than 10 million copies in 40 languages, achieving the kind of success almost unheard of in the history of science writing. His subsequent books, *The Universe in A Nutshell* and *The Future of Spacetime*, with Kip S. Thorne and others, have also been well received.

He was born in Oxford, England, on January 8, 1942 (300 years after the death of Galileo). He studied physics at University College, Oxford, received his Ph.D. in Cosmology at Cambridge, and since 1979 has held the post of Lucasian Professor of Mathematics. The chair was founded in 1663 with money left in the will of the Reverend Henry Lucas, who had been the Member of Parliament for the University. It was first held by Isaac Barrow, and then in 1663 by Isaac Newton. It is reserved for those individuals considered the most brilliant thinkers of their time.

Professor Hawking has worked on the basic laws that govern the universe. With Roger Penrose, he showed that Einstein's General Theory of Relativity implied space and time would have a beginning in the Big Bang and an end in black holes. The results indicated it was necessary to unify General Relativity with Quantum Theory, the other great scientific development of the first half of the twentieth century. One consequence of such a unification that he discovered was that black holes should not be completely black but should emit radiation and eventually disappear. Another conjecture is that the universe has no edge or boundary in imaginary time.

Stephen Hawking has twelve honorary degrees and is the recipient of many awards, medals, and prizes. He is a Fellow of the Royal Society and a Member of the United States National Academy of Sciences. He continues to combine family life (he has three children and one grandchild) and his research into theoretical physics together with an extensive program of travel and public lectures.